The Gateway to Understanding: Electrons to Waves and Beyond WORKBOOK

Matthew M. Radmanesh, Ph.D.
Professor of Electrical
& Computer Engineering,
California State University,
Northridge

authorHOUSE

1663 LIBERTY DRIVE, SUITE 200
BLOOMINGTON, INDIANA 47403
(800) 839-8640
www.authorhouse.com

First published by AuthorHouse 09/19/05

ISBN: 1-4208-3999-3 (sc)
ISBN: 1-4208-4046-0 (dj)

Library of Congress Control Number: 2005902839

Printed in the United States of America
Bloomington, Indiana

This book is printed on acid-free paper.

Cover Illustration: *The design on the cover shows the progression of concepts to application masses; a journey from the viewpoint (the sun), to postulates (the clouds), to fine energy particles (the dots), leading to waves (the radiated light), down to its condensation to solid matter (the mountains) and eventually arriving at the final end point: the panoramic physical universe in all of its aspects, spanning from the microscopic to the macroscopic universe-- ranging in size from an atom, to a device, to a circuit, to a satellite, to the earth, to the solar system, to the Milky Way galaxy and beyond.*

*Dedicated to
the Educators and
Skilled Practitioners of
Modern Sciences
Who Have Greatly Enriched
Our Lives!*

Contents

Foreword..ix

Preface...xiii

Acknowledgements....................xxi

Part One: The Essentials....................1

1 How to Learn Sciences........................3

2 Why This Course?..........................19

3 The Course Features.......................21

4   How to Use This Workbook................23

5 The Course Goals.........................27

6 The Reader's Understanding............31

7 The Gateway to Understanding:
   Course Checksheet.........................35

Certificate of Achievement............…...…...........**57**

Part Two: The Appendixes.............….. 59

Appendix A   LESSONS...........….....…........... 61
   Lesson 1: The Scientific Fundamentals............…....63
   Lesson 2: The Anatomy of the Physical Universe.69
   Lesson 3: The Workable Postulates of Physics
         and Engineering...........................75
   Lesson 4: The Discovery of Static Electricity......79
   Lesson 5: The Discovery of Kinetic Electricity....83
   Lesson 6: The Discovery of Magnetism from
         Electricity.................................87
   Lesson 7: The Discovery of Interaction: Kinetic
         Electricity with Matter...................…..91
   Lesson 8: The Discovery of Force Fields.......... .95
   Lesson 9: The Theoretical Discovery of Electronic
         Waves...................................101
   Lesson 10: The Experimental Proof of Waves 105
   Lesson 11: The Supplemental Discoveries.........109
   Lesson 12: Beyond the Modern Age...............115
   Lesson 13: Conclusion...........................121

Appendix B   ANSWERS TO QUIZZES......125
   Chapter 1 Answers................................127
   Chapter 2 Answers................................137
   Chapter 3 Answers................................149
   Chapter 4 Answers................................159
   Chapter 5 Answers................................165
   Chapter 6 Answers................................171

Chapter 7 Answers.....................................175
Chapter 8 Answers.....................................179
Chapter 9 Answers.....................................187
Chapter 10 Answers...................................197
Chapter 11 Answers...................................201
Chapter 12 Answers...................................213
Chapter 13 Answers...................................235

Appendix C   ANSWERS TO THE FINAL
            QUIZ **243**
The Final Quiz Answers..........................245

Appendix D   CLASSICAL LAWS OF
            ELECTRICITY............…..……..**253**

Glossary of Technical Terms...................257

Technical References...................…......……293

About the Author..............................…..299

Foreword

To "The Gateway to Understanding: Electrons to Waves and Beyond" Book

The study of electronic waves has been broken down into artificial subdivisions where most texts cover a small part of the subject mathematically while ignoring the rest. In our scientific community, there has been a need for a coherent , one-stop approach that covers the breadth of this material in a manner that allows a total comprehension of the subject. I believe this book finally fills this void by delivering this long-awaited material.

The field of electronic waves has not only had a profound impact on our scientific thinking and understanding of the universe around us, but has found numerous applications based on the mathematical sophistication of Maxwell's equations.

Most people in our highly automated society strive toward a meaningful understanding of the workings of electronic equipment and the major principles behind the fascinating form of energy called electricity and electronic waves. Most textbooks on the subject, however, ignore the basic concepts and introduce such a great deal of mathematical complexity that most students of science consider the subject of electricity untenable. Professor Radmanesh has written this book with the hope of inspiring and enlightening those individuals who desire an understanding of the world of electricity based on a top down approach with no undue complexity.

It will become obvious to the reader, as it has to me, that this book is the result of many years of observation and codification of the physical universe with particular attention to a deep understanding of electronic waves. In this work the origin of our universe is examined and many novel principles are freshly and elegantly introduced.

The book begins by examining the scientific fundamentals and presents a solid understanding of the material universe and its three original postulates leading to four constituent components, which have paved the way to its numerous applications (Chapters 1- 3).

Chapters 4 through 9 present a panoramic view of waves based on major discoveries that took the world of sciences by surprise when they were first introduced in the nineteenth century. Electronic waves are clearly shown to be merely an example of how waves are part and parcel of this universe. With a unique approach, this book shows how to construct and handle any universe such as a technical universe or any component thereof such as waves.

Chapter 10 presents experimental proof of existence of waves, an important step in any scientific endeavor, where anything proposed must be proven. This is a remarkable step in the discovery of waves and Dr. Radmanesh has done a superb job in providing the reader with a clear understanding of what goes into making a successful hypothesis, along with accurate methods to perform scientific experiments to prove the hypothesis.

The subject of electronic waves has been a gateway to wireless networks, cellular phones, Global positioning system (GPS), Microwave Monolithic integrated circuits (MMICs), etc. The inventions and scientific developments in this and the last century provide a glaring testimony to this.

There are Eight supplemental discoveries, which are presented in Chapter 11 with an eye to their practical applications and how they can be used in the solution of actual problems dealing with different aspects of electromagnetism. These supplemental discoveries help to obtain quick answers to many complicated technical problems facing the professionals in this fascinating field.

Chapter 12 treats the reader to a delightful discourse on the author's vision of what lies beyond the physical universe. It presents a stimulating view of what forms the make-up of any science and presents a clear view of our complex physical universe both on a classical as well as quantum mechanics level.

In Chapter 13, the author makes several concluding remarks in what he hopes will be helpful suggestions in bringing about a revitalization in thinking for scientists and engineers, and methods of education of current and future science students and/or inventors.

The basics are explained clearly and the powerful principles of electricity are expressed lucidly and dynamically, providing a keen impression in the reader's mind. It is written for technical as well as non-technical readers and should serve as a valuable resource for professional engineers, scientists, teachers, undergraduate/graduate students, interested but non-technical individuals, and technical managers.

The significance of this work lies in the manner in which it explores the fundamental postulates forming our universe and presenting them to the scientific community for the first time. Furthermore, the book makes a definite connection between the field of electricity and the broad aspects of our material universe, and does it with tremendous simplicity while making the reader aware of more intricate aspects of our sciences.

Finally, this book through its clarity and straightforward style of presentation provides the scientist as well as the non-technical individual an opportunity to appreciate the relationships between two seemingly unrelated universes: the "physical universe" and the "thought universe."

Dr. Asad M. Madni, CEng., FIEEE, FIEE, FAAAS, FNYAS, FIAE, FIBA
UCLA Alumnus of the Year Award, 2004

President & Chief Operating Officer, BEI Technologies, Inc.

Preface

To "The Gateway to Understanding: Electrons to Waves and Beyond" Book

A few years ago the author published a popular text entitled "Radio Frequency and Microwave Electronics, Illustrated," which was well received by the scientific community and at the time it appeared that some of the main goals set forth by the author in writing the book were fulfilled. However, the main public who were not technically oriented or mathematically gifted did not instantly gravitate toward this work. Thus an attempt to present and bring about a general understanding of sciences, particularly the field of electronic waves, was begun.

Almost anyone in today's highly technological society strives toward a higher understanding of the inner workings of electronic equipment and desires to know the major principles behind this fascinating form of energy called electricity and electronic waves, and yet most electrical books present the basic concepts with so much complexity and filled with so many mathematical equations that the average public individual has given up on the subject and perforce has decided to retire to the sideline to be a spectator on the subject. In other words, his hope has been dashed aside and his dream of a higher understanding has not been fulfilled in any of the modern texts on electricity.

Furthermore, the subject of waves, particularly electronic waves has been piecemealed to a point where every text covers only a small part of the subject and leaves the rest to someone else to develop. At the present moment there is no coherent and one-stop approach that covers this material from A to Z and presents it in such a way that average public can wrap their wits around it. There has been a need for such a text for quite sometime.

Over the last hundred years, the field of electronic waves, initially placed on a firm ground by Maxwell, has grown and blossomed magnificently, primarily in the area of applications to match up with the mathematical sophistication with which Maxwell's equations were presented.

The present work is the culmination of many years of study, observation, and pondering on the dilemmas and enigmas of propagating waves and their origin and the resultant understandings that was extracted from this sophisticated and at times incomprehensible field of study.

In preparing this book, the emphasis was shifted from sophisticated mathematical solutions and shifted to conceptual understanding of the material contained in the postulates and axioms of a science. Instead the complex mathematical equations have been transferred to the many appendixes in the back of the back. The emphasis has been aptly placed on workable postulates of physics and essential principles of electromagnetism and the related discoveries using simple concepts while emphasizing the basics all the way.

Throughout this book the reader will be delighted to find numerous examples of actual application mass that has been developed through the years by very bright minds and practitioners of this field. The focus is on some of the basic applications of the electromagnetics to many different fields of study including, electrical circuits, magnetic circuits, RF engineering, microwave systems, electronics and photonics.

It is an interestingly uncommon book written primarily for the technical as well as the non-technical man. It is intended to serve several classes of our society

a) The professional engineers,
b) The technical inventors,
c) The technically versed individuals,
d) The interested but non-technical individuals,
e) The college professors,
f)   The business and industrial leaders,
g) The college and university students,
h) The professional scientists, and
i)   The Non-Engineers.

The book can be used for professional and practicing Engineers in the field, and business or industrial leaders, who are the visionaries of their group. It can also be used for inventors of new devices and gadgets. Most inventors would like to go to a source, where the initial spark of invention can be ignited. This book will surely serve as well this important class of our society—the technical Inventors.

For the average man who may or may not be technically versed and yet desires to learn about the universe at large, or the technical world in his immediate surroundings. It is intended to lift the aura of "black magic" surrounding the world of sciences particularly electricity, to enlighten and demystify the subject of sciences in the minds of ordinary individuals.

The Importance of Work

Rather than looking into the complicated mathematical equations for solutions, Man's long search for answers to the riddles of the technical world will finally be amply rewarded through the pages of this book. By avoiding undue complexities, the reader will achieve occam's razor doctrine and will be actually traveling in the direction of "the actual why" and thus be able to put his thoughts on the right track for all the future problems forthcoming.

Within the confines of this book, one is given a chance for the first time to take an in-depth look and inspect firsthand, one of the most dynamic fields of study that has ever been developed in the history of mankind on this planet. The basics are laid in simple terms and clear explanations express the powerful principles lucidly and dynamically, providing an unforgettable impression in the reader's mind.

The scientist, armed with the tools solidly laid out in this book, will be well equipped to understand scientific journals and handle the problems of work-a-day world of sciences, particularly testing, analysis, and design of devices, circuits, and systems dealing with electric, magnetic or electronic wave phenomena. The increased depth of knowledge will allow one to achieve one's objectives with a much higher probability of success in this rapidly advancing subject.

The broad importance of this work could be summed up as a totally new approach to understanding our scientific world through the use of newly discovered fundamentals (missing in all technical books), which add a tremendous amount of simplicity and clarity to very complex problems.

This is a new approach unmatched in any extant text today. The discovery of these fundamentals has had a huge impact on our current world and has truly made our scientific arena a bright beacon of hope with a renewed interest in understanding our physical universe. This work has created a "unified theory "about the physical universe in very simple terms.

Finally, this work paves the way for the scientist as well as the non-technical individual to formulate and develop a relationship between two distinct universes: the "Material or Physical Universe" and the "Universe of Thought".

The Scope of Work

The current work can be used to serve as a textbook for a course in physics, electromagnetics, RF electronics, microwave Engineering and the advancing field of photonics. It is intended for all levels of college particularly for senior-level or graduate students. It can just as well serve as an excellent reference guide for the professional scientist, practicing engineer, whether in electromagnetics, RF and microwaves, electro-optics, or general sciences who is actively involved in his/her respective field.

The book starts from the very general postulates, considerations and laws and presents an example of these principles with the introduction and evolution of the universe of electronic waves. The book, divided into four parts and 13 chapters, presents these chapters with the progressive development of concepts following the same pattern as presented in the pyramid of knowledge in Chapter 1, which is:

A) PART I- THE PHILOSOPHICAL FOUNDATION OF SCIENCES

Chapters 1-3: Present a framework for any science and analyze the physical universe by putting forth the three original postulates and the derived primary postulates that created it. Moreover, the workable postulates of physics are discussed in depth and through introduction of the hidden and implicit postulates, the potential pitfalls of the subject are exposed.

B) PART II- ELECTRONS TO WAVES: THE MONUMENTAL DISCOVERIES

Chapters 4-6: Present the progressive development of electricity and magnetism and state the three monumental discoveries that brought them to the forefront of technology.

Chapters 7-9: Deal with the final saga of electromagnetism as a unified science, as it was discovered and how it came to its full fruition.

Chapter 10: Introduces the experimental proof of waves and shows conclusively the existence of waves. This was the crowning moment for Maxwell and his wave theory, who predicted the existence of waves long before its discovery.

Chapter 11: Introduces the supplemental discoveries, which help to fill the gaps of knowledge in electromagnetism and to elucidate those subordinate concepts, which were not touched upon in the previous chapters.

c) PART III- THE FUTURE AHEAD

Chapters 12-13: Discuss what lie beyond the current universe and presents a summary of the basic scientific considerations that underlie all of our current sciences and technology. It also points out to the universe ahead and what would the future bring at the current pace of scientific developments.

d) PART IV-APPENDICES

A list of symbols used in each chapter and a series of quizzes are included at the end of each chapter to aid the reader gain a fuller understanding of the presented materials. The book ends with a glossary of technical terms and several important appendices. These appendices cover physical constants and other important data needed in the process of understanding of the material covered in the text.

The Author's Goals

As part of the author's goals, the presentation of a number of scientific fundamentals pours the foundation for understanding our complex universe particularly the universe of waves. They provide a rather deep philosophical bed-rock upon which the weight of the rest of the ensuing complicated concepts can be supported.

Moreover, with the help of this foundation, the author intends to achieve several milestone achievements:

a) *Create a better perspective toward application of sciences especially electricity and electronic waves,*

b) *Promote a better and deeper understanding of the scientific methodology, and how it is used to achieve stellar results in the scientific arena,*

c) *Bring about a public awareness of sciences and how the universe of waves affects us in many ways on a constant basis and how a simple knowledge of the subject leads to a plethora of applications.*

The full "pyramid of knowledge," pertaining to the field of waves, has been streamlined and presented in a simple language. This missing element (pyramid of knowledge) from the field of science and engineering as well as education has created a truncated science where the top half of this pyramid is cut off. This vital amount of information missing from any subject invites disaster for its students and followers. It would surely create great turmoil in the minds of people associated with sciences (e.g. students, educators, designer, etc.) all over the world. The brief presentation of these basics in this chapter is expected to bring about nothing but a small renaissance in the scientific arena as a whole particularly in the field of waves, electricity and magnetism which are rapidly advancing fields.

The tools used by the pioneers and founders of sciences are the tools of today's successful scientists, who strive to succeed in their profession. By knowing these materials well, the reader will be taking the same route that all the great men and women of sciences took to achieve their remarkable discoveries that we are all benefiting from.

The non-technical reader will be invited to examine a series of basic materials that will enable him/her to understand his immediate universe far better than ever before. He/she will be exposed to materials of considerable significance, which surely would open up the gates of knowledge along with a wider horizon of understanding.

For professional engineers or science students who are armed with these tools as their stock-in-trade, they are equipped well to understand scientific journals and handle the problems of work-a-day world of sciences, particularly analysis and design of devices, circuits and systems dealing with electric, magnetic, electronic waves, etc. Their depth of knowledge will allow them to achieve their objectives with a much higher probability of success in this magnificent arena.

Matthew M. Radmanesh, Ph.D.

xx

Acknowledgements

First and foremost, special thanks are due to Dr. L. Ron Hubbard, whose writings, whether fiction or non-fiction, inspired the author to write the original manuscript of the book and then the current workbook, which makes learning more practical. It helped to ignite the initial spark of realization of how sciences connect with the universe of thought and how they are actually a smaller subset of a much larger sphere of activity called life and livingness. Thus, a quest for the answers of a puzzle of enormous proportion, i.e. electricity, was begun.

The author also wishes to thank many of his undergraduate and graduate students spanning twenty years (1984-2004), who with their inquisitive minds, propelled the author to revise the first manuscript of this work in order to answer some of their questions.

Moreover, amongst many the author is grateful to Jeff Quin of Litton Guidance Center whose assistance in many ways was invaluable. Special thanks to Dr. Myungkul Kim, a great microwave student and friend, who made many invaluable suggestions to the first manuscript, to Larry Alpers for many valuable discussions and words of encouragement when everything looked insurmountable. Thanks are also due to Andrea Hom for superb graphics work and highly professional illustrations.

The author's special gratitude also goes to two of his greatest friends, Robb and Tami Gray, who helped him in many ways to keep his thoughts focused and provided the much needed support during this intense project. Thanks are also due to two highly-valued and special individuals, Jaime and Linda Rodriguez, who simplified many challenging aspects of this work.

The author would like further to thank many of his professional colleagues particularly Dr. Asad Madni (CEO/COO of BEI Technologies, Inc.), one of his best friends/colleagues, John Guarrera, a great human being, Dr. S. T. Mau (Engineering Dean), Dr. E. S. Gillespie, Dr. Ichiro Hashimoto, Dr. Nagi El Naga, Dr. Jeff Wiegley, Dr. John Noga, Dr. Ramin Roosta, Dr. Tim Fox, Dr. Robert Burger, Dr. Tom Mincer, Ben Mallard (California State University, Northridge, CA), dear friends at CSUN, Philip Arnold (HP), a good friend, Dr. George Haddad & Dr. C. M. Chu (University of Michigan, Ann Arbor, MI), the early mentors in Michigan, Dr. M. Torfeh and Dr. Robert Salajekeh (Kettering University, Flint, MI), two good friends/associates. Their support and collegiality through the years is definitely appreciated.

Finally, author's deep gratitude belongs to his lovely wife, Jane Marie, his ever-playful son, Jeremy William, for making life soothing and sweet during this power-packed project and to his parents, Mary and the late Dr. G. H. Radmanesh, for their true love and unconditional support.

Matthew M. Radmanesh, Ph.D.
Dept. of Electrical and Computer Engineering,
California State University, Northridge,
April 2005

Part One

The Essentials

1

How to Learn Sciences

THE IMPORTANCE OF CORRECT EDUCATION
Education in the science of physics, engineering, electronics and others consists of guiding the reader along a gradient of known data, with the highest attention to the basic concepts that form the foundation of any of these desired fields of study.

The basic concepts presented in this book are far more fundamental than the mother sciences of engineering (i.e., physics and mathematics) and cover the essential truth about our physical universe in which we live. These basic truths convey a much deeper understanding about the nature of the physical universe than has ever been discussed in any electronic or for that matter any scientific textbook.

These basic truths set up a background of discovered knowledge by mankind, against which a smaller sphere of information (i.e., electricity, electronic engineering, physics, etc,) can be examined. Many of the principles, which appear in technical books, are easily describable and thus understood much better once the basic underlying concepts are grasped.

While studying sciences and engineering at the university, the author always looked for simplicity, a higher level of truth, and a deeper level of understanding in all of the rigorous mathematics and many of the physical laws that were presented. Upon further investigation, the underlying principles, which form the backbone of all extant physical sciences, have finally emerged and are presented as the "fundamentals of physical sciences".

In general, we can see that workable knowledge is like a pyramid, where from a handful of common denominators efficiently expressed by a series of basic postulates, axioms and natural laws, which form the foundation of a science, an almost innumerable number of devices, circuits, and systems can be thought up and developed.

The plethora of the mass of devices, circuits and systems generated is known as the application mass, which practically approaches infinity in sheer number. This is an important point to grasp, since the foundation portion never changes (a static) while the base area of the pyramid is an ever changing and evolving arena (a kinetic), where this evolution is in terms of new implementation techniques and technologies.

The emphasis has been shifted from rigorous and sophisticated mathematical solutions of complex equations and instead has been aptly placed on the conceptual understanding of electronic waves by using simple explanations, while emphasizing the basics all the way. This is the use of occam's razor in actual practice, which strips away all complexities in sciences and brings forth the simplicity of thought and observation.

The emphasis is on fundamentals and for this reason all new technical terms must be thoroughly defined and understood as they are introduced. This is a novel approach and is based upon the results obtained in recent investigations and research in the field of education, which has shown that the lack of (or the slightest uncertainty on) the definition of terms will pose as formidable obstacles in the reader's mind in achieving full comprehension of the material.

A series of uncomprehended or misunderstood technical terms will block one's road to total comprehension and mastery of the subject. This undesirable condition will eventually lead to a dislike and total abandonment of the subject.

Therefore, the road before one is at once a clear-cut path to knowledge by understanding that one needs to grasp the terms fully before one grasps the concept, and all of these should occur long before one achieves mastery of any desired subject.

FUNDAMENTALS

The materials presented in this section deals with the fundamental considerations of science and engineering. These fundamental concepts have been ignored, neglected and even omitted from most scientific texts much to the detriment of the students. This section presents a first-hand glimpse of these vital scientific basics and attempts to enhance the depth of understanding of physical sciences in general, and electricity in particular.

KNOWLEDGE AND SCIENCE, DEFINITIONS OF

Since we are embarking upon a road to knowledge, it is imperative that we define it at the outset of this work:

Definition-Knowledge: *A body of facts, principles, data and conclusions (aligned or unaligned) on a subject, accumulated through years of research and investigation which provides answers and solutions in that subject.*

This is an important definition since it lays the groundwork for what comes ahead in the future. As students of knowledge, we must understand that every time we are studying a subject, we are traveling on the road to knowledge (see Figure 1).

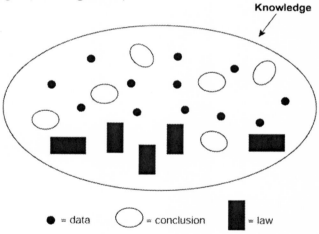

FIGURE 1 Definition of knowledge.

Therefore, the source of information and the presenter's perspective and depth of knowledge on the subject plays an important role in one's learning and guidance along this road to knowledge. This point should not be taken lightly, since lack of a good and reliable source of information and sometimes the presence of a false source, is the downfall of many bright minds who, with proper guidance, would have otherwise been very productive and brilliant in that field of study.

In our modern age we are surrounded with many fields of study each with its own sphere of knowledge. The most common approach to understanding a subject, as presented at the modern universities today, is through a technical vehicle called a "science", which is defined as:

Definition- Science: *A branch of study concerned with establishing, systematizing and aligning laws, facts, principles and methods which are derived from hypotheses, observation, study and experiments.*

From these two definitions we can see that in contrast to knowledge, a science is an aligned body of data, facts and principles based upon natural laws which have similarity in application (see Figure 2).

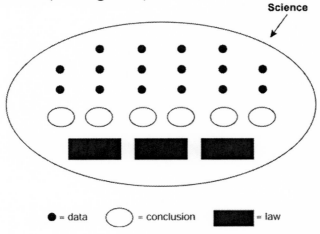

FIGURE 2 Definition of a science.

STRUCTURE OF A SCIENCE

We can observe that any science can be roughly divided into two divisions:

a) <u>**Considerations**</u>: This division is an extremely important part of any science and primarily deals with all the theories, facts, research findings, postulates, hypotheses, design methodologies, technology, manufacturing techniques, discovered principles of operation, natural laws, etc.

b) <u>**Application Mass**</u>: This division deals with all the related masses which are connected and/or obtained as a result of the application of the science. This includes all physical devices, machines, experimental set-ups and other physical materials that are directly or indirectly derived as a result of the application of the science. For example for the field of electrical engineering, this division includes all electrical devices, circuits, motors, machinery, transmission lines, antennas, computers, assemblies, subsystems, systems, networks, etc., that are designed or directly obtained utilizing the electrical engineering theory and principles.

It is vital to note that of the two divisions, the former has much higher importance and seniority over the latter, even though the latter is much more visible, voluminous and more abundant in quantity. Therefore this work's primary focus is on the considerations related to the field of electricity and the discussion on application masses connected with the subject are all relegated to manufacturer's handbooks, datasheets, etc.

CONSIDERATIONS BUILT INTO A SCIENCE

In the field of considerations, we can observe that not every piece of data has the same level of importance as others. In fact there are a few which rank the highest than others and form the foundation of the science. In the order of importance, the major considerations in any science can be summarized as follows:

a) Fundamental scientific postulates
b) Axioms and Natural laws
c) Nomenclature
d) Fundamental theorems
e) Analysis and theory of operation
f) Charts, diagrams, tables, etc.
g) Design methodology and procedures
h) Technology
i) Manufacturing techniques,
j) Computer analysis and application software design methodology, considerations, etc.

Due to the importance of these terms, and their repeated use we will now define a few of them briefly below as:

Definition- Scientific Postulate: *A technical statement that requires no proof, and is assumed to be true unconditionally and for all times.*

Definition- Natural Laws: *A body of workable principles considered as derived solely from reason and study of nature.*

Definition- Theorem: *A proposition which is not self-evident but can be proven from accepted premises, and so is established as a principle.*

Definition- Technology: *The application of knowledge toward practical ends; the technical materials dealing with the know-how of producing a product or a specific result.*

FOUNDATION OF A SCIENCE

From the viewpoint of considerations, any extant science has a foundation upon which it is built. This is very similar to a building's physical structure where its footing and foundation is of essential importance, since it supports the weight of all the upper floors and its contents. The foundation of a science consists of two parts:

a) The fundamental postulates
b) Axioms and Natural laws

Upon this solid foundation rests all theoretical research, extrapolations and design methodologies, application mass and all future explorations, inventions and discoveries. The natural laws are discovered by observation and study but nevertheless have a lot in common with the fundamental postulates of a science.

It can be observed that the fundamental postulates of a science form the bedrock upon which all natural laws rest. This means that the postulates and the discovered natural laws together, form the foundation of a science. It is an important concept which is omitted in the majority of scientific texts.

All the remaining considerations such as scientific conclusions, technical data, design methods, rules, and the entire application mass of the subject rest on top of the foundation as shown in Figure 3.

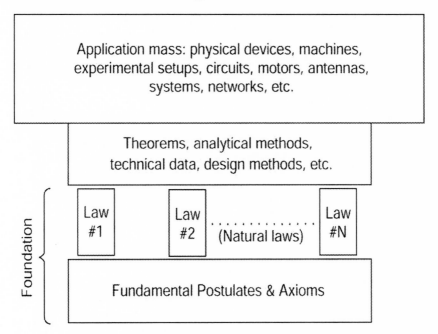

Figure 3 The organization of the electrical science.

We can summarize our observations into a useful conclusion:

Conclusion: *The closer the laws of a science are to simplicity and the basic postulates, the higher the workability of that science.*

THE PYRAMID OF KNOWLEDGE

As discussed in the Gateway book, practical knowledge is highly hierarchical and can be cast into the form of a pyramid, called the "Pyramid of Knowledge". The pyramid of knowledge in a workable science has been developed and is shown in Figure 4.

From this pyramid we can see that the postulating viewpoint lies at the top followed by the Fundamental Postulates, the Axioms and Natural Laws (at the apex) forming the foundation part of any science, whereas the Nomenclature followed by Theorems, Theory & Analysis, Charts, Tables & Diagrams, Design methodology, Application Technology down to manufacturing principles create the fundamental concepts. At the base of the pyramid, there is the application software and the application mass as the eventual end point of the postulates of that science.

Figure 4 can now be used as a fundamental and yet a very accurate template to model any science after, in order to build it from the ground up. In this manner, one can understand the basic structure of any extant physical science or develop the basic framework for any future new science.

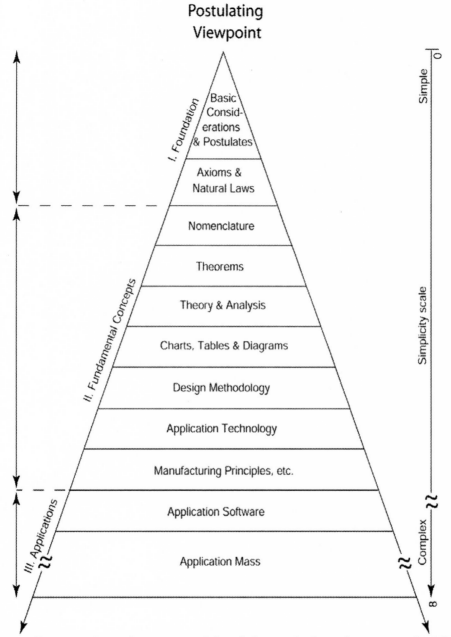

FIGURE 4   The pyramid of knowledge in a workable science.

For example, the pyramid of knowledge in the vital field of electrical engineering, leading to the subjects of "Radio Frequency Electronics", "Microwave Electronics", "Photonics", "Electro-Optics", "X-Ray Electronics", etc., has been developed and is shown in Figure 5.

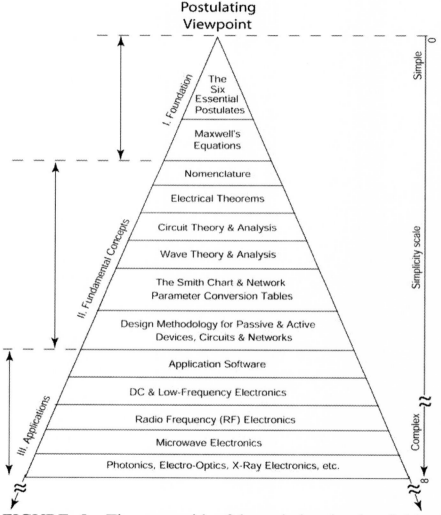

FIGURE 5   The pyramid of knowledge in the field of electrical engineering.

Using the pyramid of knowledge is a valuable way to organize a science and thus achieve a higher understanding and a greater simplicity in solving or analyzing complex problems.

Furthermore, by observation we can generalize the pyramid of knowledge to include all of the extant sciences. This generalization would take the shape of a circle. we can see the entire field of knowledge of mankind about the physical universe takes on the form of a circle or a pie, where each workable science is a slice of the pie. At the center of the circle, we place the fundamental postulates of the physical universe which are held in common by all of the sciences. This common area, at the center of the circle, forms an inter-relatedness amongst all sciences (see Figure 6).

AXIOMS AND PRINCIPLES

There are a series of exact axioms and principles, associated with studying sciences, which need to be mastered before one can grasp the full extent of scientific world.

These basic axioms and principles are covered in depth in the Gateway book but a brief listing of them may help to bring about a better alignment of thought and understanding of the complicated world of sciences. These are:

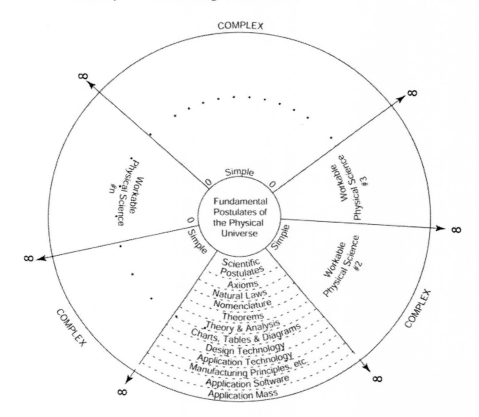

FIGURE 6   The circle of knowledge interrelating all workable sciences.

A) **BASIC AXIOMS**
1) The Uniqueness Axiom
2) The Mutual-Exclusiveness Axiom
3) The Non-Absoluteness Axiom
4) The Nomenclature Axiom
5) The Application Mass Axiom

B) **BASIC PRINCIPLES**
1) The Principle of Relativity of Knowledge
2) The Universal Communication Principle

3) The Uncertainty Principle
4) The Exclusion Principle
5) The Principle of Bilateral Communication
6) The Dichotomy Principle
7) The Duality Principle

For example, using the pyramid of knowledge in combination with these basic axioms and principles, the whole field of electricity has been organized in a top-down fashion and its many laws (and their duals) have been developed as shown in appendix D. From this appendix, we can see that the field of electricity can be understood much better if one uses these basic concepts as organizing tools, i.e., as the common denominator of knowledge, to simplify and present all of the discovered laws and/or scientific principle more effectively.

SCIENTIFIC TERMINOLOGY

It could be observed that knowledge or information on a subject, when put into a language form becomes quantized. In other words knowledge is not continuous but rather made up of information quanta.

The basic building blocks (or quanta) of information of any science, are its technical terms or terminology. Each technical term (or quantum of information) carries with it an exact package of information which is embodied in its definition. The set of specific technical terms, which is an inseparable part of any scientific subject, is also commonly referred to as its nomenclature.

This concept indicates that in any topic, or generally in any body of knowledge or any scientific study, specific and exact definitions of terms are necessary in order to comprehend and communicate to others the laws, observations, assumptions, problems, solutions and other relevant facts and conclusions.

Moreover, it is a well-known fact in the education arena that having an inadequate comprehension of the terminology is one of the leading causes of confusion and misunderstanding of the subject. Therefore, the terminology of a science forms an important part of the subject and mastery of any subject requires mastery of its terminology along with its accurate definitions.

This means that a student, when faced with a stream of information regarding a scientific concept, needs to fully grasp each quantum of information (i.e. each technical term), with the exact definition that was originally intended by its author, before the whole scientific concept can be grasped.

2

Why This Course?

This self-study course was prepared in response to an overwhelming demand for a course, which could more effectively utilize the popular book entitled *"The Gateway to Understanding: Electrons to Waves and Beyond,"* (also called the Gateway Book) published by AuthorHouse, 2005.

The Gateway book presents the fusion of science with philosophy as a unified theory, for the first time in the history of sciences. This has never been done before and thus is an important point to grasp.

The Gateway book starts from very general postulates, considerations and laws and chapter by chapter narrows the focus down to very specific concepts and applications, culminating in the design of various electronic circuits. Toward the end of the book, sciences are shown to be a subset of life and livingness, and their connection to a much larger subject, i.e., philosophy is amply demonstrated.

This workbook follows the same pattern as set forth by the Gateway book which is written on a top-down approach, that is to say, the most important considerations, that are true at all times, are presented first and gradually the scope is narrowed down to very specific applications. This approach is extremely powerful because it allows the reader to have "the whole picture" before he focuses on a "small part" of it.

The initial motivation was to bring the basics to the forefront and orient the reader in such a way that he/she can think with these fundamentals correctly. This concept led to developing this course, covering the thirteen chapters in the Gateway book.

3

The Course Features

Skills or Knowledge Acquired

After finishing this course, you should be able to have the following knowledge skills:

1) Understand the fundamental concepts underlying the science of electricity, particularly electronic waves and electromagnetism.
2) Have a good understanding of important electrical quantities, their exact definitions with their corresponding units.
3) Know the most fundamental electronic laws and their limitations.
4) Know the postulates underlying the field of electricity and the universe of waves.
5) Understand the physical universe better and be able to operate in it more successfully.

Advantages

Learning the subject of electricity through the study of this course is tremendously more beneficial than simply purchasing and reading the book on your own. The course provides many advantages including:

a) A step by step approach presenting a series of lessons, which are bite-sized pieces of information taken from the book.

b) The lessons act like a trail or a "road to knowledge" with an exact beginning and a finite end. This prevents possible frustration of the reader from aimlessly reading the book or getting overwhelmed by the enormity of the subject.

c) Solutions to many of the end of chapter problems provide an excellent check-out to the reader's comprehension of the material.

d) A streamlined approach to the learning process, which takes irrelevant materials off the direct path of achieving the final goal of the course: total comprehension of electricity and its many aspects.

e) Author's numerous comments, exercises and summary adds clarity and understanding and brings simplification to a very complicated subject.

Intended Audience

The course is intended for the practicing engineer or any individual who desires a workable knowledge and intuitive understanding of electricity, magnetism, electronic waves and the considerations that lie beyond. The course is taught from a very practical point of view and the use of higher mathematics is minimized. It is a highly recommended course for any practicing engineer, manager, or lay person who would like to gain a deeper understanding as well as a broader knowledge of electricity in connection with the thought universe.

4

How to Use this Workbook

The comprehensive course, as presented in this workbook, is studied step by step through the use of a checksheet as will be discussed in section 7 of this workbook.

For each chapter in the Gateway book, we have allocated a section in the checksheet, which covers that chapter in depth (i.e., sections I-XIII). At the end of each section of the checksheet, there is a "Homework" assignment, which is based upon lessons provided in the appendix A.

Appendix A is divided into thirteen lessons. Each lesson is divided into several sections identified by an icon. The icons for each section are different and each closely resembles the concept presented in that section.

The lessons in appendix A were developed with the intent that the reader would adhere to the following study strategy:

1)

 **LEARNING OBJECTIVES**

At the beginning of each lesson, the Learning Objective will describe the goals of that particular lesson. This section enables you to know the concepts that will be presented and treated in that lesson.

2)

 **READING ASSIGNMENT**

In this section, there will be an assigned section of the text that the reader needs to read before he will be able to proceed with the rest of the lesson.

3)

 **KEY POINTS**

This section provides important highlights of what was read in the "reading assignment" section.

4)

 **PRACTICE PROBLEMS**

This section provides a number of problems that the student needs to solve based on the material in that lesson. These problems are designed to test your comprehension, and reinforce the materials in that lesson by using specific

sections of the textbook. You learn better by applying the "learned concepts" to "practical problems".

5)

ANSWERS TO PRACTICE PROBLEMS

The answers to "Practice Problems" are in the Appendix B of this workbook.

6)

SUMMARY

This section presents a brief summary of the lesson and makes some concluding comments that will add a final touch to the material studied.

Within each of the thirteen lessons, you will be asked to answer the end of chapter quizzes that are presented in the Gateway book. The detailed answers to these quizzes can be found in Appendix B.

There is a final exam in section XIV of the checksheet, which is based on the final quiz as presented at the end of chapter 13 of the Gateway book. The detailed answers to the final quiz can be found in Appendix C.

THE HONOR CODE: *All of the quizzes and the final exam are taken under the "Honor Code" defined as:*
"I will neither give nor receive any unauthorized information from any unapproved source." This means that one may use the relevant parts of the Gateway book to answer the questions in the quizzes, but can not use the "Answers to Quizzes" <u>while</u> *taking the quizzes.*

However, one is allowed and actually encouraged to use the "Answers to Quizzes" to correct one's quiz, only <u>after</u> the quiz (or the final exam) has been taken; this is the true purpose of the "Answers to Quizzes", and provides a correction action, which is vital in the learning process!

Finally upon successful completion of this course, in section XV of the checksheet, you will be asked to attest to the completion of this course. This is a remarkable achievement on your part, which regardless of the obstacles facing you while taking this course, you still managed to take it to its final conclusion! Furthermore, At this point of the checksheet, you can fill out the "Certificate of Achievement", detach and keep it for future reference. You can also write to the author and announce your completion in writing.

Therefore, you are hereby congratulated, in advance, for your perseverance culminating in the successful completion of this course.

5

The Course Goals

The presentation of a number of scientific fundamentals pours the foundation for understanding our complex universe particularly the universe of waves. They provide a rather deep philosophical bed-rock upon which the weight of the rest of the ensuing complicated concepts can be supported.

Moreover, with the help of this foundation, the workbook intends to achieve several milestone achievements:
   a) Create a better perspective toward the application of sciences, especially electricity and electronic waves,
   b) Promote a better and deeper understanding of the scientific methodology, and how it is used to achieve stellar results in the scientific arena, and
   c) Bring about a higher appreciation of sciences and how the universe of waves affects us in many ways on a constant basis and how a simple knowledge of the subject leads to a plethora of applications.
   d) Bring about a general understanding of why physical sciences have taken such a dominant role in our society and what led to their appalling

success, which has greatly outpaced the "Social Sciences", such as psychology, sociology, humanities, psychiatry, etc.

NOTE: *It is interesting to note that the "Social Sciences", with their lackluster level of understanding of the human species, have bypassed their main goal of "Understanding Human Beings" and have shifted their focus to the animal kingdom (e.g., rats, monkeys, etc.) and actually intend to achieve their goal by animal studies! It should become gradually evident to the reader that while these are still floundering in their infancy of development, the physical sciences have become a towering giant, with the resultant harnessing of enormous forces of nature, which is put under the control of Man, who has yet to mature into a fully responsible being, and to take charge of the destiny of himself as a species, his progeny and the physical universe with all of its inhabitants as a whole. Such a lack of maturity in responsibility for Man directly and squarely falls on the shoulder of our "infant social sciences", which despite elevating themselves to the level of "sciences" have no concourse with scientific methods of analysis, etc., and have yet to achieve the precision results or the success of their physical counterpart.*

e) Make a clearer connection between "Life and Livingness" and "Sciences" and describe how the former creates the latter, as shown in Figure 7.

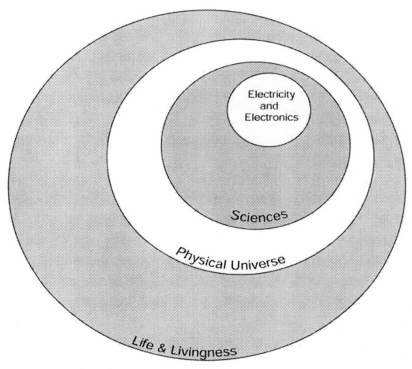

FIGURE 7 The field of "Electricity and Electronics" in correct perspective with regard to other fields.

This workbook utilizes the many discovered principles, particularly the pyramid of knowledge, where it has been streamlined and presented in a simple language in the Gateway book.

Moreover, the missing parts of the pyramid of knowledge in sciences and/or engineering, observable in the higher institutes of learning and classrooms all around the Globe, has created truncated subjects where the top half of this pyramid is cut off. This vital amount of information missing from any subject invites disaster for its students and followers. It would surely create great turmoil in the minds of people associated with sciences (e.g., students, educators, designers, etc.) all over the world, and thus goes on the confusing state of affairs in sciences!

The brief presentation of these basics in this work is expected to bring about nothing but a small renaissance in the scientific arena as a whole, particularly in the field of waves, electricity, and magnetism, which are rapidly advancing fields.

The tools used by the pioneers and founders of sciences are the tools of today's successful scientists, who strive to succeed in their profession. By knowing these materials well, the reader will be taking the same route that all the great men and women of sciences took to achieve their remarkable discoveries that we are all benefiting from.

Professional engineers or science students who are armed with these tools as their stock-in-trade, are well-equipped to understand scientific journals and handle the problems of work-a-day world of sciences, particularly analysis and design of devices, circuits and systems dealing with electric, magnetic, electronic waves, etc. Their depth of knowledge will increase and thus will allow them to achieve their objectives with a much higher probability of success in this magnificent arena.

The non-technical reader will be invited to examine a series of basic materials that will enable him to understand his immediate universe far better than ever before. One will be exposed to materials of considerable significance, which surely would open up the gates of knowledge along with a wider horizon of understanding.

6

The Reader's Understanding

The reader's understanding plays a major factor in his/her progress in this course. Thus one must use self discipline to stay abreast of the materials at all times. This is done by clearing up areas of confusion or misunderstandings through the use of the Gateway book glossary, or better yet, a technical dictionary. Use of these tools is essential for successful course comprehension.

Learning is not so much an additive process, with new materials simply piling up on top of the existing knowledge, but an active and dynamic process in which the mental connections are constantly changing.

Ideally speaking, the information filing system or the way knowledge is stored in the mind of a student who is well prepared with all fundamentals in, is in the form of a multidimensional tree. Real learning would be in terms of a constant updating and changing of this multi-dimensional structure (at the top) as the new data arrives, with the fundamentals forming the root of the structure (at the bottom), staying unchanged or perhaps being strengthened as new fundamentals are learned.

Excitement of learning comes when new connections are formed, either a) by creating brand new ones sitting on top of prior connections, or b) by pulling apart some old connections and replacing them with fresh ones. The important point to grasp here is that the new information results in meaningful learning only when it connects with what already exists in the mind of the learner!

The learning of a novice is labored and slow, not because the novice is less intelligent, or less motivated than an expert, but because the connections between the new information and the existing mental structure are sparse, and moreover there are no mental hooks on which to hang the new information on and thus it falls in a heap on the floor in a confused state, recalled with great difficulty.

In other words, we need to understand the extreme importance of a sound mental foundation and the vital role that it plays to prepare one in learning or understanding sciences, or in general any desired subject, without which study or learning becomes impossible.

The outlined series of steps in this workbook, if done exactly as prescribed, would undoubtedly pour a new mental foundation or would further strengthen an existing one. This would not only prepare one to learn sciences in a new unit of time, but also would lead to an enrichment of one's life with the newly learned information.

The eventual goal of the process of learning is empowering one to reach greater heights of understanding and opening many doors of opportunity to fulfill many dreams, which would have been otherwise unrealized.

Furthermore, it brings about a more enlightened life, which would make this world a much more pleasant place to live in.

The author considers it one of the most rewarding things to have another individual grasp the materials in all of its simplicity and realize the true purpose of sciences, particularly electricity, engineering and physics, and then use it intelligently and ethically to increase own as well as mankind's potential survival in this universe and help others to achieve their goals and in the process, make Man take control of his own destiny, without being shackled by the chains of higher authority or superstition.

Moreover, it is the author's basic philosophy of life and one of his greatest joys, to be able to extend a hand of friendship and help to others and let them share their understandings and realizations while studying or doing the various exercises in this workbook. Therefore, any communications in the way of a healthy criticism and/or correction in order to improve the quality of this work, are greatly welcome and should be sent directly to:

Dr. Matthew M. Radmanesh
18111 Nordhoff Street,
Department of Electrical and Computer Engineering,
California State University, Northridge, California 91330.
Or email to: **matt@csun.edu**

7

The Gateway to Understanding: Course Checksheet

Date started:_____

Student's name:_____

1. The Purpose of a Student

A student by definition is one who studies. He is an attentive and systematic observer, and reads in detail in order to learn and then to apply. A student through reading, observing and applying is traveling on a road to knowledge, so as he studies this subject, he should know that his purpose is to understand the materials he is studying and not memorization. He should connect what he is studying now to what he will be doing in actual practice, at a later time.

2. How to Use the Checksheet

There are certain specific actions that a student needs to perform in order to get this checksheet done as follows.

a. *Checksheet Sequence*: This checksheet lists, in sequential order, the actions you should do to complete the course. Therefore, in doing this checksheet, please make sure that you do not jump around but you follow it sequentially, that is, you do each step of the checksheet in the order that is presented.

b. *Sign-off:* The space to the right of each item is underlined and is intended for your "initials and date" when you have completed the action as directed. Example: *Initials/date MR-9/10/04.*

c. *Essays and Practicals:* After the completion of each of these steps, write down your observations and conclusions on a separate sheet of paper and then sign off that item on your checksheet.

d. *Sketches and Plots:* Use of sketches or plots is highly recommended while studying the materials in the Gateway book and throughout each phase of this course. It adds mass, doingness and involvement to the ideas and concepts presented in the text. These sketches, plots, or drawings bring a new level of clarity and comprehension to the concepts and ideas, since these can be seen and visualized.

e. *Textbook*: The textbook to be used throughout this course is:
M. M. Radmanesh *"The Gateway to Understanding: Electrons to Waves and Beyond"*, AuthorHouse, 2005. For brevity, this text is referred to as the "Gateway book" in this workbook.

3. The Prerequisites

A basic desire to understand scientific concepts as well as a strong need and curiosity to gain a deeper comprehension of scientific methodology and/or its basic philosophy of operation, are required and highly recommended.

4. The Course Objectives

The objectives of this course are: a) to make you able to understand how and why electricity works, b) understand the basic fundamentals behind our technological society, and c) to train you how to analyze problems concerning electronic waves or electricity expertly.

SECTION I- FUNDAMENTAL CONCEPTS
(Chapter 1)

1.1 The Gateway book: Study sections 1.1-1.5.

_____

1.2 Essay: a) Describe the generalized concept of the universe.

_____

b) Describe what a scientific postulate is and why it is important.

_____

1.3 What is a scientific postulate and why is it important? Write down your answers.

_____

1.4 Sketch the pyramid of knowledge in electrical engineering.

_____

1.5 The Gateway book: Study sections 1.6-1.12.

_____

1.6 Essay: a) Write down why knowing nomenclature is essential.

_____

b) Describe what is meant by the physical universe dichotomy as well as the duality principle and their application to the engineering sciences.

_____

c) Sketch how you would solve any complex problem.

_____

d) Write down the role of mathematics in sciences.

_____

e) What is meant by the generalized concept of relativity?   _____

f) What is the differences between physical sciences and social sciences? Draw diagrams.   _____

1.7 Homework: Do Lesson 1 as given in appendix A.

_____

SECTION II- THE ANATOMY OF THE PHYSICAL UNIVERSE
(Chapter 2)

2.1 The Gateway book: Study sections 2.1-2.5.

_____

2.2 Essay: a) What is meant by the "Big Picture" and how do we apply it to sciences.   _____

b) Describe in your own words, the four primary postulates.   _____

a) Describe what is meant by space and curved space.

_____

2.3 The Gateway book: Study section 2.6.

_____

2.4 Essay: Write down the definitions of objective time and subjective time.   _____

2.5 Sketch a diagram showing how time can become nonlinear.   _____

2.6 The Gateway book: Study sections 2.7-2.8.

_____

2.7 Essay: a) Describe and define what is meant by the original postulates.   _____

b) Write down what is meant by the missing vital element and how it creates space. _____

2.8 Sketch the generalized concept of viewpoint.

_____

2.9 The Gateway book: Study section 2.9-2.12.

_____

2.10 Essay: Describe a) what are the basic axioms and principles used in sciences. _____

b) Write down the list of properties of a postulate versus an application mass. _____

2.11 Homework: Do lesson 2 in appendix A.

_____

SECTION III- THE WORKABLE POSTULATES OF PHYSICS AND ENGINEERING
(Chapter 3)

3.1 The Gateway book: Study sections 3.1-3.5.

_____

3.2 Essay: a) Describe the two workable postulates of physics. _____

b) Describe the two implicit postulates of physics.

_____

c) Write down what is the hidden element and why is it important. _____

c) Describe how a coordinate system and a viewpoint relate to each other. Give an example. _____

3.3 Sketch a coordinate system and show the origin as well as an observing viewpoint. _____

3.4 The Gateway book: Study sections 3.6-3.8.

_____

3.5 Essay: a) Write down the six hidden postulates and describe each with an example.   _____

b) Describe what is meant by the particle-wave duality.

_____

c) The Heisenberg Uncertainty principle and its ramifications in measurement on a large or small scale.

_____

d) The particle-wave duality and how it applies to the dual theory of light.   _____

3.6 Practical #1: Redo example 3.2 with Q_1 the same but Q_2=-2x10^{-19} C, m_2=10^{-30} kg.   _____

(Try varying the charge values and see the effect on the results).   _____

3.7 Homework: Do lesson 3 in appendix A.

_____

SECTION IV- THE DISCOVERY OF STATIC ELECTRICITY
(Chapter 4)

4.1 The Gateway book: Study sections 4.1-4.6.

_____

4.2 Essay: a) Write down the definition of electric charge.

_____

b) Describe what is meant by the evolution of thought in electricity.   _____

c) Describe the Coulomb's electric force law._____

4.3 Sketch a diagram showing all aspects of what the first monumental discovery has led to.   _____

4.4 Practical: Examine static electricity by doing a simple experiment such as by rubbing two things together (such as hands, feet, etc.) and measuring the added charge by a very sensitive instrument such as a micro-voltmeter or a lie detector; alternately, if such instruments are not readily available you may want to examine a picture or a diagram of static electricity such as lightning, a charged capacitor, etc. and then write down your observations.

_____

4.5 The Gateway book: Study sections 4.7-4.8.

_____

4.6 Essay: Explain a) How does a copy machine work?

_____

b) How does a lightning rod work?

_____

c) How does a paint spraying machine work?

_____

d) How does a Cathode Ray Tube work?

_____

e) How does an electric wheel work?

_____

4.7 Practical #1: a) Go to the real world and examine the electrostatic application mass in your environment. Write down your observations.

_____

b) Using your imagination, write down some examples of your own concerning possible application mass in electrostatics.

_____

4.8 Exercise: (optional) Summarize and write a brief description for solved examples L.1-L.3 in section I of appendix L of the Gateway book.

_____

(Try varying some of the numbers and see the effect on the results).

_____

4.9 Homework: Do lesson 4 in appendix A.

_____

SECTION V- THE DISCOVERY OF KINETIC ELECTRICITY
(Chapter 5)

5.1 The Gateway book: Study sections 5.1-5.3.

_____

5.2 Essay: a) Write down the definition of kinetic electricity.

_____

b) Write down Galvani's observation and why Volta opposed Galvani's theory.

_____

5.3 Essay: a) Describe what is meant by an essential milestone in electricity.

_____

b) Describe how Volta made the first battery.

_____

c) What is the principle of operation of a battery?

_____

d) What is the definition of electromotive force? Explain.

_____

5.4 Diagram: Draw a diagram showing the evolution of electrokinetics, from discovery to application mass through time.

_____

5.5 The Gateway book: Study sections 5.4-5.6.

_____

5.6 Essay: a) Write down the concept of emf generation.

_____

5.7 Essay: a) Explain the nine methods of emf generation briefly.

_____

b) What is a condenser microphone and how does it work?

_____

c) Write down one possible application of electrokinetics from your own observation. _____

5.8 Exercise: a) Observe one example of electrokinetics in your environment. Write down your observation.

_____

b) (optional) Exercise: Summarize and write a brief description for solved example L.4 in section II of Appendix L in the Gateway book. _____

5.9 Homework: Do Lesson 5 in appendix A.

_____

SECTION VI- THE DISCOVERY OF MAGNETISM FROM ELECTRICITY
(Chapter 6)

6.1 The Gateway book: Study sections 6.1-6.4.

_____

6.2 Essay: a) Write down what is meant by a magnet.

_____

b) Write down the two types of magnets and describe each.

_____

c) What was the magical discovery by Oersted?

_____

6.3 Essay: a) Describe the two new uses for a magnetic needle. _____

b) Write down a few examples, from your own observation, about the uses of a magnetic needle.

_____

6.4 The Gateway book: Study sections 6.5, 6.6.

_____

6.5 Essay: a) Describe in your own words the concept of magnetostatics.   _____

b) Describe magnetic disk memory as an application mass.

_____

c) What is a magneplane? Explain.

_____

d) What is a rail gun? Explain.

_____

6.6 Sketch a diagram showing what the discovery of magnets has led to.   _____

6.7 (optional) Exercise: Summarize and write a brief description for solved examples L.5, L.6 and examples L.1 and L.2 in section III of Appendix L in the Gateway book.

_____

(Try varying some of the numbers and see the effect on the results).   _____
6.8 Homework: Do lesson 6 in appendix A.

_____

SECTION VII- THE DISCOVERY OF INTERACTION: KINETIC ELECTRICITY WITH MATTER
(Chapter 7)

7.1 The Gateway book: Study sections 7.1-7.4.

_____

7.2 Essay: a) Write down what is meant by electro-dynamics.   _____

b) Describe why the interaction is an important discovery.

_____

7.3 Sketch: Draw a diagram to show the relationship between a current and the corresponding magnetic field.

_____

7.4 Describe a) why electricity is more senior to magnetism?

_____

b) What is meant by Ampere's circuital law?

_____

7.5 The Gateway book: Study sections 7.5 and 7.6

_____

7.6 Essay: a) Describe, in your own words, how the first alternator worked.

_____

b) Write down the difference between a dynamo and an alternator.

_____

c) Write down how a cyclotron and an isotope separator work.

_____

d) Describe how the first electromagnetic motor worked.

_____

7.7 Practical: a) examine a small battery-operated motor of a toy or of a similar apparatus. Write down your observations.

_____

b) (optional) Summarize and write a brief description for solved examples L.7, L.8, and example L.3 in section IV of Appendix L in the Gateway book.

_____

7.8 Homework: Do lesson 7 in appendix A.

_____

SECTION VIII- THE DISCOVERY OF FORCE FIELDS
(Chapter 8)

8.1 The Gateway book: Study sections 8.1-8.8.

_____

8.2 Essay: Write down what is meant by the concept of a force field.

_____

b) What was the essence of Newtonian Theory?

_____

c) What was the revolutionary concept in electricity?

_____

d) What is the universal nature of force (or energy)?

e) What are the lines of force and how was it used to reveal conservation of charge?

_____

8.3 Practical: Using a world Atlas examine the earth's magnetic field and draw the lines of force around earth.

_____

8.4 The Gateway book: Study sections 8.9 and 8.10.

_____

8.5 Essay: Describe, in your own words, the operation of:
    a) A transformer,

_____

    b) The earth's ionosphere,

_____

    c) A grating reflector,

_____

    d) A frequency meter,

_____

    e) A doubled communication capacity using the two- polarization technique.

_____

8.6 (optional) Exercise: Summarize and write a brief description for solved examples L.9-L.10 in section V of Appendix L in the Gateway book. _____

8.7 Homework: Do lesson 8 in appendix A.

_____

SECTION IX- THE THEORETICAL DISCOVERY OF ELECTRONIC WAVES
(Chapter 9)

9.1 The Gateway book: Study sections 9.1-9.9.

_____

9.2 Essay: a) Explain why electronic wave is the final piece of the puzzle. _____

b) What is the nature of waves? _____

c) What was the Maxwell's contribution? _____

d) What were the original 8 and the reduced 4 Maxwell's Equations? _____

e) Explain what is meant by displacement current.

_____

9.3 Sketch: Draw a diagram showing how Maxwell's equations lead to RF/Microwaves and optics.

_____

9.4 The Gateway book: Study sections 9.10 and 9.11.

_____

9.5 Sketch: a) Draw a diagram showing how charge is the smallest subset of the universe of waves. _____

b) Draw a diagram showing the sunlight's different types of polarizations.

_____

9.6 Essay: a) Write down one example of RF applications.

_____

b) Describe one example of Microwave applications.

_____

c) Describe one example of Optical applications.

_____

9.7 (optional) Exercise: Summarize and write a brief description for solved examples L.11-L.13, and example L.4 in section VI of Appendix L in the Gateway book.

_____

9.8 Homework: Do lesson 9 in appendix A.

_____

SECTION X- THE EXPERIMENTAL PROOF OF WAVES
(Chapter 10)

10.1 The Gateway book: Study sections 10.1-10.11.

_____

10.2 Essay: a) What is Maxwell's legacy?

_____

b) Describe the four properties of electromagnetic waves.

_____

c) Describe the importance of discovery of electrons and photons.

_____

d) What was Hertz's observation?

_____

10.3 Sketch: a) Draw the Hertz apparatus for demonstrating the speed of waves in air and wire.

_____

b) Draw a diagram showing the AC power distribution proposed by Tesla.

_____

c) Draw the fields of study in electrical engineering.

_____

10.4 Essay: a) Describe the shortcomings of Edison's DC power distribution system.   _____
b) Explain why Tesla's AC power system won over Edison's DC system.   _____
c) Explain why optics is a subset of electromagnetic waves.   _____
10.5 Essay: Write down why Hertz's experimental proof was an essential step in accepting Maxwell's electromagnetic theory.   _____
10.6 Homework: Do lesson 10 in appendix A.

_____

SECTION XI- THE SUPPLEMENTAL DISCOVERIES
(Chapter 11)

11.1 The Gateway book: Study sections 11.1-11.6.

_____

11.2 Essay: Write down what is meant by:
    a) Supplemental discoveries,   _____
    b) The first supplemental discovery and its magnetic dual,   _____
    c) The second supplemental discovery and its magnetic dual,   _____
    d) The third supplemental discovery and its magnetic dual,   _____
    e) The fourth supplemental discovery,

_____

    f) The fifth supplemental discovery.

_____

11.3 Essay: Write down what is the definition and meaning of:

 a) Corollary 5a,   _____

 b) Corollary 5b,   _____

 c) A superconductor.   _____

11.4 Essay: Describe what is the electronic structure of matter and how does it bring about electric current flow.

             _____

11.5 The Gateway book: Study sections 11.7-11.11.

             _____

11.6 Essay: Write down what is meant by:

 a) The sixth supplemental discovery and its magnetic dual.   _____

 b) The seventh supplemental discovery and its magnetic dual.   _____

 c) The eighth supplemental discovery and its magnetic dual.   _____

11.7 Sketch: a) Draw the current flow in a conductor showing electron's motion.   _____

b) Draw a diagram summarizing all of the supplemental discoveries and the resulting application mass.

             _____

11.8 Homework: Do lesson 11 in appendix A.

             _____

SECTION XII- BEYOND THE MODERN AGE
(Chapter 12)

12.1 The Gateway book: Study sections 12.1-12.6.

             _____

12.2 Essay: Write down and explain what is meant by:

    a) The role that postulates play in sciences.

    _____

    b) The main postulates of the physical universe.

    _____

    c) The common denominator of postulates.

    _____

    d) The exact sequence of postulates in the physical universe. _____

    e) The generalized application mass versus considerations. _____

12.3 Sketch: Draw a diagram showing the exact sequence of postulates: from space to galaxies. _____

12.4 Essay: Write down what are the two major errors in the field of medicine and what would cause the subject to fail someday. _____

12.5 The Gateway book: Study sections 12.7-12.16.

    _____

12.6 Essay: Write down and explain what is meant by:

    a) The three categories of application mass.

    _____

    b) The collective viewpoint of physics.

    _____

    c) The generalized viewpoint. _____

    d) Purposes lying beyond postulates. _____

12.7 Sketch: a) Draw a diagram summarizing the discoveries in electricity. _____

b) Draw a diagram showing the finiteness of the physical universe and how it can appear to be infinite in size.

    _____

Essay: Write down and explain what is meant by:

a) Nano-technology. _____

b) The universe ahead of us. _____

12.8 Homework: Do lesson 12 in appendix A.

_____

SECTION XIII- CONCLUSION
(Chapter 13)

13.1 The Gateway book: Study sections 13.1-13.5.

_____

13.2 Essay: Write down and explain:
a) The primary factors in the construction of the physical universe.

_____

b) Why there is apparent certainty at the classical level but at the quantum level there is uncertainty.

_____

c) The nature of the macroscopic universe.

_____

d) The characteristics of the microscopic universe.

_____

e) The three irreducible postulates in the physical universe.

_____

13.3 Sketch: Draw a diagram showing the scientific arena in relationship to the microscopic and macroscopic universes and the vital element.

_____

13.4 Essay: a) Explain the six salient points built in the construction of any science, or in general any universe.

_____

b) Describe what the "mindset of God" was when the universe was made and what it means to us now.

_____

13.5 Sketch: Draw a diagram showing the relationship between the life force and an application mass.

_____

13.6 Practical: a) Go outside and examine the three original postulates and the four primary postulates in actual action in nature and life. Write down your observations. _____

b) Examine a few postulates in your own personal life and how it led to its personalized application mass. Now discover what were your purpose(s) behind those postulates. Write down your observations.

_____

13.7 Homework: Do Lesson 13 in appendix A.

_____

SECTION XIV- FINAL EXAM

14.1 Take the "Final Quiz" as given at the end of Chapter 13 of the Gateway book. (This is a self test and one must use the "Honor Code" while taking this test).

_____

14.2 Grade your Final Quiz using the answers given in appendix C.

(Grade your quiz by assigning two point for each totally correct answer, and one point for each partially correct answer. The answers do not need to be verbatim but should convey the correct concepts).

_____

a) If score is 85% or above, you have passed this course. Congratulations and jump to section XV! _____
Otherwise, go to step (b).

b) Restudy the uncomprehended sections or areas of difficulty and retake the quiz without looking at the answers and then regrade your answers. _____

c) If the result of step (b) is 85% or above go to section XV, otherwise repeat steps (b) and (c) until you genuinely pass this exam.   _____

SECTION XV- COURSE COMPLETION AND ATTESTATION

To the student:

I attest that I have gone fully through all of the assigned materials for this course, have done all practicals, examples, sketches, exercises, quizzes, homework, etc. and have fully completed all items as called forth in the checksheet. I have no doubts, reservations or uncertainties about the materials presented in this course. My name and signature below indicates my full compliance with the checksheet requirements:

Student name:_____

Student signature: _____

Completion Date: _____

Upon attestation, please fill in your name and the date of completion in the blank area in the certificate of achievement on the next page, detach, and keep for your future reference.

Congratulations!

End of Checksheet

Certificate of Achievement

It is Hereby Certified That

_____
(Name)

Has Satisfactorily Attained and Completed All the Necessary Requirements and is Awarded the Certificate of Achievement for the Completion of

"The Gateway to Understanding: Electrons to Waves and Beyond" *Course*

Matthew M. Radmanesh, Ph.D.
Course Founder

Date_____

Part Two

The Appendixes

Appendix A

LESSONS

LESSON 1

The Scientific Fundamentals

 LEARNING OBJECTIVES

After completing this lesson, you will be able to:

1 Demonstrate an understanding of the foundation of electricity.

2 Recognize the postulates of physics and electrical engineering.

3 Identify the main reasons why definition of terms and understanding the technical nomenclature is important.

4 Analyze and observe the dichotomies and dualities in the physical universe.

READING ASSIGNMENT

Review the following sections of Chapter 1 of the book:
Sections 1.1-1.12.
These sections discuss the fundamental concepts of science and engineering.

KEY POINTS

1. Any science is roughly divided into two divisions: a) considerations, and b) application mass.

2. Every workable science is built upon a handful of powerful postulates from which all natural laws can be derived.

3. The postulates and natural laws of any science form the foundation of that science.

4. The application mass (i.e. all of the devices, circuits and systems) of electrical engineering and their behavior or response is a byproduct of the fundamental concepts of this science and not vice versa.

5. Electrical engineering as a field of study is formed like a pyramid where there are only four

postulates at the apex (having a total mass of zero) and a plethora of devices, circuits and systems (approaching an infinite mass) at the base of the pyramid. This is what we mean when we say, *"from zero to infinity"*.

6. Nomenclature of electricity is an essential and inseparable part of the science and mastery of the terms and their accurate definitions are a must for successful course comprehension.

7. Everything in electricity comes in duals or "opposite pairs". If one learns about one item of a pair, the second item (the opposite item) can be understood with considerable ease.

8. To solve any problem we need to: a) Dissect the problem, b) Use natural laws, c) Obtain a first order solution (or baseline solution), and d) Fine tune the solution iteratively, and finally e) obtain the final answer (or the nth-order solution).

9. Introduction of arbitrary factors or assumptions in the solution process, which are unjustified or unverified, will always bring about a worse solution and thus a wrong answer.

PRACTICE PROBLEMS

Do Quizzes: 1.1 through 1.3.
(This is a self test and one must use the "Honor Code" while taking this test).

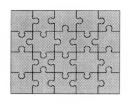

ANSWERS TO PRACTICE PROBLEMS

See "Appendix B: Answers to Quizzes, Chapter 1"

(Use appendix B to grade your quizzes by assigning two point for each totally correct answer, and one point for each partially correct answer. The answers do not need to be verbatim but should convey the correct concepts. If the score is below 85%, restudy the chapter and retake the quiz; If need be repeat this process until the score is above 85% and only then continue to the next step).

SUMMARY

Lesson 1- Summary
To master electricity and electronics, one needs to do the following:

1. Fully comprehend fundamental natural laws of electricity.

2. Master exact definitions of basic technical or electric terms.

3. Learn a systematic method to solve any engineering problem.

4. Learn the basics so that one can understand and do analysis or design various electronic circuits with relative ease.

5. Practice these techniques under the guidance of an expert until a full mastery of the subject is achieved.

LESSON 2

Anatomy of the Physical Universe

LEARNING OBJECTIVES

After completing this lesson, you will be able to:

1 Know the fundamental concepts and the primary postulates forming the physical universe.

2 Gain an understanding of space, energy, matter, and time as the basic components making up the physical universe.

3 Know exactly the meaning and definition of basic axioms concerning our universe.

4 Understand the dichotomy of postulate and application mass.

5 The importance and meaning of postulate zero, the most vital element in all of sciences.

 **READING ASSIGNMENT**

Review the following sections of Chapter 2 of the book:
Sections 2.1-2.12.
These sections discuss the fundamental concepts regarding the physical universe.

 **KEY POINTS**

1. The basic building blocks in the physical universe are based upon four primary postulates.

2. There are three original postulates from which everything in the physical universe is made.

3. There are four fundamental components in the physical universe:
    a) Linear Space.
    b) Waves and fine energy particles.
    c) Condensed light energy particles leading to matter particles and objects.
    d) Monotonically increasing "Mechanical time."

4. There is a dichotomy in space: "created space-created space."

5. The concept of dichotomy can be extended and generalized to energy, matter, and time, exposing the true nature of the physical universe as a universe of created spaces, created energies, created masses and automatically created time.

6. Engineering analysis and design is a constant interplay between the two sides of the generalized dichotomies (create/created) of space, energy, matter and time.

7. The seven basic axioms and principles sets up a solid background of information against which any smaller area of interest, such as electricity or electronic weaves, can be examined with tremendous success to create a host of sciences, with electricity as their common denominator.

 **PRACTICE PROBLEMS**

Do Quizzes: 2.1 through 2.4.
(This is a self test and one must use the "Honor Code" while taking this test).

ANSWERS TO PRACTICE PROBLEMS

See "Appendix B: Answers to Quizzes, Chapter 2)

(Use appendix B to grade your quizzes by assigning two point for each totally correct answer, and one point for each partially correct answer. The answers do not need to be verbatim but should convey the correct concepts. If the score is below 85%, restudy the chapter and retake the quiz; If need be repeat this process until the score is above 85% and only then continue to the next step).

SUMMARY

Lesson 2- Summary

1. The physical universe as an application mass is founded upon a few basic but exact postulates or concepts. These concepts in the order of importance can be summarized as:

a. The phenomenon of space is senior to all other postulates.

b. The postulation of energy occurs only in an already created space.

c. The consideration of condensing the finer energy particles into heavy particles of matter leads to matter and solid objects.

d. The concept of interaction of matter with energy on a continuous basis and at a fixed rate sets up the basis for the establishment of mechanical time and its units, and thus allowing the measurement of all other motions in the physical universe.

2. The physical universe is an incomplete universe, in that, it can not understand itself or communicate answers for its own resolution.

3. The physical universe is an "effect type of universe", and the creation or byproduct of a viewpoint (quite visible on a small scale) and thus needs an exterior viewpoint to solve its mysteries or complexities.

4. The dichotomy principle as well as the duality principle are great tools in analyzing any universe particularly the physical universe, and are extensively used in all aspects of engineering.

LESSON 3

The Workable Postulates of Physics and Engineering

 **LEARNING OBJECTIVES**

After completing this lesson, you will be able to:

1 Understand the two interwoven universes of physics.

2 Recognize what the workable postulates of physics are and their differences with the primary postulates.

3 Understand the hidden postulates of physics and why they are important.

4 Comprehend the particle-wave duality.

5 Understand the dual theory of light.

READING ASSIGNMENT

Review the following sections of Chapter 3 of the textbook: Sections 3.1-3.8.
These sections discuss the basics of physics and its application in electrical engineering.

KEY POINTS

1. The first workable postulate is the fragmented forms of energy, which includes such things as current, voltage, power, work, force, etc.

2. There are two implicit postulates in physics, which are taken for granted by most scientists:
> a) Implicit postulate of created space, which is linear and has infinite dimensions.
> b) Implicit postulate of mechanical time, which is monotonically increasing on a linear basis.

3. A Coordinate system is related to a viewpoint but they are not equivalent to each other and not even comparable. A viewpoint in physics is far superior to a coordinate system.

4. Particles and waves are duals of each other, and form the woof and warp of our physical universe.

5. Duality of a light wave and a photon is an example of particle-wave duality.

6. The Uncertainty principle is inherent in the physical universe and shows that one can never achieve a 100% certainty in the measurement of any quantity. This means that absolutes are impossible to achieve in the physical universe.

PRACTICE PROBLEMS

Do Quizzes 3.1 through 3.4.
(This is a self test and one must use the "Honor Code" while taking this test).

ANSWERS TO PRACTICE PROBLEMS

See "Appendix B: Answers to Quizzes, Chapter 3)
(Use appendix B to grade your quizzes by assigning two point for each totally correct answer, and one point for each partially correct answer. The answers do not need to be verbatim but should convey

the correct concepts. If the score is below 85%, restudy the chapter and retake the quiz; If need be repeat this process until the score is above 85% and only then continue to the next step).

 **SUMMARY**

Lesson 3- Summary
The science of physics has certain built-in limitations which need to be understood clearly, in order to master sciences in the true sense of the word. The highlights of these facts can be summarized as:

1. Microscopic universe and macroscopic universe have different concepts for energy.

2. The implicit postulates are an important part of physics and their exact understanding is just as important as the main workable postulates of physics.

3. Knowledge of the hidden postulates is important because it allows one to see the underlying principles that have been operating in physics for centuries. Lack of awareness of them acts like a landmine, which can trip an unsuspecting student of the subject or even the advanced practitioners, without ever realizing the insidious role that they play in our modern sciences.

LESSON 4

The Discovery of Static Electricity

 **LEARNING OBJECTIVES**

After completing this lesson, you will be able to:

1 Understand the evolution of thought in electricity.

2 Recognize the significance of the first discovery.

3 Understand what is meant by electric charge and the two types of charge.

4 Know how to generate electric charge.

5 Know the meaning of Coulomb's force law.

READING ASSIGNMENT

Review the following sections of Chapter 4 of the textbook: Sections 4.1-4.8.
These sections discuss the basic concepts in static charge

KEY POINTS

1. The evolution of thought has played a big role in the discovery of electricity.

2. The first discovery put the subject of electricity on a very firm and solid ground

3. The codification and measurement of charge turned the field of electricity from a totally qualitative subject to a quantitative science and catapulted the subject into a whole new arena of scientific endeavors.

4. Coulomb's force law expressed clearly the factors that the electric forces between static charges depend on. The electrostatic force is:
   a) Proportional to the value of each charge.
   b) Inversely proportional to the distance between the two charges.

5. The discovered principles in electrostatics has led to the development of enormous application mass over the last four centuries.

PRACTICE PROBLEMS

Do quizzes 4.1 through 4.3.
(This is a self test and one must use the "Honor Code" while taking this test).

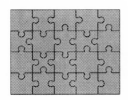

ANSWERS TO PRACTICE PROBLEMS

See "Appendix B: Answers to Quizzes, Chapter 4"

(Use appendix B to grade your quizzes by assigning two point for each totally correct answer, and one point for each partially correct answer. The answers do not need to be verbatim but should convey the correct concepts. If the score is below 85%, restudy the chapter and retake the quiz; If need be repeat this process until the score is above 85% and only then continue to the next step).

 **SUMMARY**

Lesson 4- Summary

The science of electrostatics is based upon very specific laws and rules. The highlight of this lesson can be summarized as:

1. Electric charge has a specific definition with certain exact properties.

2. There are two types of charge: positive and negative, which are dual of each other.

3. Formulating the force law between static charges and further codification of electrostatics has led to the development of many devices, circuits and machines that have served mankind magnificently through the years and have rewarded him amply for his discovery of this energy form.

4. Electricity, not necessarily in the static form, is constantly and continually being utilized by life organisms on a daily basis on a mental as well as physical level, in their effort to survive on this planet.

LESSON 5

The Discovery of Kinetic Electricity

 **LEARNING OBJECTIVES**

After completing this lesson, you will be able to:

1 Understand the essential milestone, which changed the course of electricity forever.

2 Recognize the definitions of electromotive force (emf).

3 Recognize the methods of generation of emf.

4 Understand the source of bimetallic electricity.

5 Know how to solve problems by applying electrokinetics to obtain solutions or application mass.

READING ASSIGNMENT

Review the following sections of Chapter 5 of the textbook: Sections 5.1-5.6.
These sections discuss the fundamentals of kinetic electricity.

KEY POINTS

1. The discovery of kinetic electricity was an essential milestone in the progress of electricity toward a mature subject.

2. Galvani's observation of the frog leg's trembling was the spark that marked the beginning of this remarkable discovery.

3. Volta's fascinating power of observation and realization of the existence of the bimetallic electricity was the cornerstone upon which the whole field of electrokinetics is founded.

4. The invention of the first battery led to the most primary step in the evolution of electric circuits and their modern theory.

5. There are a total of nine methods of generation of electromotive force, of which electrochemical method is a significant method of producing electricity at the moment, and includes the bimetallic electricity as discovered by Volta.

6. Electrokinetics has opened up a wide open field of study, which is beyond limits from the point of view of application mass.

PRACTICE PROBLEMS

Do quizzes 5.1 and 5.2.
(This is a self test and one must use the "Honor Code" while taking this test).

ANSWERS TO PRACTICE PROBLEMS

See "Appendix B: Answers to Quizzes, Chapter 5"

(Use appendix B to grade your quizzes by assigning two point for each totally correct answer, and one point for each partially correct answer. The answers do not need to be verbatim but should convey the correct concepts. If the score is below 85%, restudy the chapter and retake the quiz; If need be repeat this process until the score is above 85% and only then continue to the next step).

 **SUMMARY**

Lesson 5- Summary
The following important points can be brought forth as the highlights of this lesson:

1. Kinetic electricity is the dual of static electricity, very much like kinetic energy, which is the dual of potential energy.

2. The prime mover in moving the electron and thus producing a current is the electromotive force (emf).

3. There are a total of nine methods of producing emf, each one valid in generating a voltage. Electrochemical is the first method used to produce electricity by means of a battery, however, electromagnetic method is currently the most popular method as will be discussed in a later lesson.

LESSON 6

The Discovery of Magnetism from Electricity

 **LEARNING OBJECTIVES**

After completing this lesson, you will be able to:

1 Understand how kinetic electricity creates magnetism.

2 Realize how magnetism is the dual of kinetic electricity.

3 Recognize the two types of magnets.

4 Recognize how a magnetic needle can be used for information transmission as well as measurement of electric current.

5 Understand how the discovery of magnetism and its connection to electricity opened a whole new vista for the physicists.

READING ASSIGNMENT

Review the following sections of Chapter 6 of the textbook: Sections 6.1-6.6.
These sections discuss the fundamental concepts behind magnetism.

KEY POINTS

1. Discovery of kinetic electricity was instrumental in discovering the separate field of magnetism to be linked to electricity.

2. There are two types of magnets:
   a) Natural magnets, which is a piece of lodestone, and
   b) Artificial magnets, which function just like a natural magnet, are obtained through three methods: 1) by stroking or rubbing, 2) by contacting through proximity (but not touching), and 3) by using a coil of current-carrying wire and winding it around the sample.
3. The link between magnetism and electricity led to measurement of a current by means of a magnetic needle and since any form of energy can be converted to a current form, this opened up a whole host of

applications never imagined before, but now they could be realized.

4. Transformation of information via a transmission line beyond the line of sight, for the first time became a reality. This was the beginning of the communication era via electrical signals.

5. Based on the information that was derived from the link between electricity and magnetism a new series of application mass were developed.

 **PRACTICE PROBLEMS**

Do Quizzes 6.1 and 6.2.
(This is a self test and one must use the "Honor Code" while taking this test).

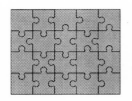 **ANSWERS TO PRACTICE PROBLEMS**

See "Appendix B: Answers to Quizzes, Chapter 6"
(Use appendix B to grade your quizzes by assigning two point for each totally correct answer, and one point for each partially correct answer. The answers do not need to be verbatim but should convey the correct concepts. If the score is below 85%, restudy the chapter

and retake the quiz; If need be repeat this process until the score is above 85% and only then continue to the next step).

SUMMARY

Lesson 6- Summary
The following important points can be brought forth as the highlights of this lesson:

1. Magnetism is a powerful force, which appears to be separate from static electricity. However, the moment the static electric charges are made to move under the force of an electromotive force, the phenomena of magnetism comes to life and becomes inseparable from kinetic electricity.

2. Using magnetism in conjunction with electricity is essential in many applications, particularly in the field of communication and mensuration.

3. Artificial magnets have an advantage over the natural magnets, and that is their magnetic property is not fixed but can be controlled externally and thus varied via, for example, an external current through the winding, etc.

LESSON 7

The Discovery of Interaction: Kinetic Electricity with Matter

 **LEARNING OBJECTIVES**

After completing this lesson, you will be able to:

1 Understand the significance of interaction of matter with kinetic electricity.

2 Understand the meaning of Ampere's force equation.

3 Recognize the seniority of electricity over magnetism.

4 Understand the Ampere's circuital law.

5 Gain the knowledge of the interaction, which can be used to bring about a plethora of application mass.

READING ASSIGNMENT

Review the following sections of Chapter 7 of the textbook: Sections 7.1-7.6.
These sections discuss the interaction of electricity and matter.

KEY POINTS

1. The discovery of the interaction of electricity with matter set forth the foundation for understanding the modern applications such as dynamos, alternators, etc.

2. Ampere's force equation clearly states that the any current carrying wire generates a magnetic field at right angles to the wire and whose magnitude is inversely proportional to the square of distance from the wire. In other words, the magnetic field follows the inverse square law.

3. The discovery of the interaction established a whole new of study called "electrodynamics."

4. Ampere showed that kinetic electricity is senior to magnetism and as such was the prime factor behind the electromagnetic fields, soon to be discovered

5. Ampere's circuital law made the integration of magnetic field around any closed loop equivalent to the current enclosed by the loop.

6. The field of electrodynamics, which dealt with applied force to matter, brought forth motion and thus made a host of applications a reality. This was a concept totally new at the time but has had huge ramifications in many aspects of our lives from electric engines, to car alternators, to space born applications, to power generators, so on and so forth.

 **PRACTICE PROBLEMS**

Do Quizzes 7.1 and 7.2.
(This is a self test and one must use the "Honor Code" while taking this test).

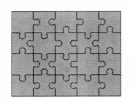 **ANSWERS TO PRACTICE PROBLEMS**

See "Appendix B: Answers to Quizzes, Chapter 7"
(Use appendix B to grade your quizzes by assigning two point for each totally correct answer, and one point for each partially correct answer. The answers do not need to be verbatim but should convey the correct concepts. If the score is below 85%, restudy the chapter

and retake the quiz; If need be repeat this process until the score is above 85% and only then continue to the next step).

 **SUMMARY**

Lesson 7- Summary
The following important points are the most important points in this lesson:

1. There is a definite interaction between moving electricity and conducting materials.

2. The interaction follows exact laws and produces a force which is proportional to the current, length of wire and the magnetic field.

3. Ampere's work showed that magnetism is a byproduct of electricity and thus junior in importance.

4. Ampere's circuital law shows a way to calculate the magnetic field caused by any given current, especially when there is symmetry in the geometry of the problem.

5. Use of principles of electrodynamics has enabled man to move matter with considerable ease and with greater efficiency.

LESSON 8

The Discovery of Force Fields

 LEARNING OBJECTIVES

After completing this lesson, you will be able to:

1 Understand the importance of force fields.

2 Know why the Newtonian theory was inadequate and erroneous to use when it comes to understanding electricity.

3 Know what the Newtonian misconceptions were.

4 Understand the concept of fields.

5 Recognize the universal nature of force and how one form of force (or energy) can be converted into another using this principle.

6 Understand how lines of force describe a force field.

7 Know the various types of force fields that exist.

READING ASSIGNMENT

Review the following sections of Chapter 8 of the textbook: Sections 8.1-8.10.
These sections discuss the concept of force fields and its applications.

KEY POINTS

1. Field is an invisible entity that represents the distribution of energy in space and is the intermediary agent in the interactions of matter with energy.

2. Newtonian concept of forces between two mass points, which assumed forces apply in a straight line and instantly could not be applied to electricity, and Faraday was the first to point this out.

3. Faraday discovered that all forms of energy are equivalent to each other and actually are various forms of a fundamental force. Thus matter, electricity, and light are of the same origin and can be converted from one form into another.

4. Lines of force are imaginary lines in a force field, which show the direction and intensity of the force fields at each point in space pictorially.

5. The "line of force" theory established the fact that the lines of the force field emanate from positive charges and terminate on negative charges.

6. The principle of conservation of charge is based upon the "line of force" theory and states that charge can not be destroyed nor created but only separated. Thus absolute charge does not exist, a fact well known from Chapter one (see section 1.6.4, The Principle of Relativity of Knowledge).

7. Seeing the truth in Faraday's force field theory, Maxwell used it as the cornerstone of his electromagnetic theory of waves.

8. There are many types of force fields but in the field of electricity, there are three types of fields: a) electric field (E-field), b) magnetic field (H-field), and c) electromagnetic field, which is a simultaneous combination of electric and magnetic fields in a special package with a certain spatial relationship between E- and H-field vectors.

 **PRACTICE PROBLEMS**

Do Quizzes 8.1 through 8.4.

(This is a self test and one must use the "Honor Code" while taking this test).

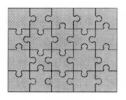

ANSWERS TO PRACTICE PROBLEMS

See "Appendix B: Answers to Quizzes, Chapter 8"

(Use appendix B to grade your quizzes by assigning two point for each totally correct answer, and one point for each partially correct answer. The answers do not need to be verbatim but should convey the correct concepts. If the score is below 85%, restudy the chapter and retake the quiz; If need be repeat this process until the score is above 85% and only then continue to the next step).

SUMMARY

Lesson 8- Summary
The following important points are the highlights of this lesson:

1. Understanding that the force field theory is essential to understanding electricity.

2. All energy forms originate from a basic source and have a common denominator, and thus have a basal unity in nature.

3. The conservation of charge is a direct byproduct of the force field theory.

4. There are three main types of fields in electricity: Electric, Magnetic and Electromagnetic fields.

5. The concept of fields has revolutionized all other subjects and has permeated many other areas totally unrelated to electricity such as chemistry, gravity, fluid mechanics, mental phenomena, etc.

LESSON 9

The Theoretical Discovery of Electronic Waves

 **LEARNING OBJECTIVES**

After completing this lesson, you will be able to:

1 Understand why electronic waves are the final piece of the puzzle in electricity.

2 Understand the nature of waves.

3 Understand the eight original Maxwell's equations.

4 Know how Maxwell's equations were modified and reduced to four.

5 Understand how the addition of the displacement current term led to the theoretical discovery of waves and the formulation of the wave equations.

READING ASSIGNMENT

Review the following sections of Chapter 9 of the textbook: Sections 9.1-9.11.

These sections discuss the fundamental concepts of waves and their discovery on paper, purely as a mathematical conjecture.

KEY POINTS

1. Waves are electronic vibrations that propagate without giving the medium, as a whole, any permanent displacement.

2. Electronic waves are a small subset of kinetic energy and include such energy forms as light waves, radio waves, and microwaves.

3. The unified theory of electricity was proposed by Maxwell, by using: a) the physical lines of force, b) a new kind of current called "displacement current", and c) a hypothetical vibrating source that propagated at the speed of light

4. Maxwell summarized his work in eight equations, wherein he expressed the relationship between

electric field, magnetic field, electric charge, and displacement current to space and time.

5. Maxwell showed mathematically that light and all of its forms (e.g., infra-red, ultra-violet, X-ray, etc.) are electromagnetic (EM) waves and propagate according to EM laws.

6. The reverse process of generating electricity from magnetism was demonstrated using a magnet in relative motion to a conductor.

PRACTICE PROBLEMS

Do Quizzes 9.1 through 9.4.
(This is a self test and one must use the "Honor Code" while taking this test).

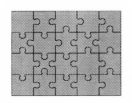

ANSWERS TO PRACTICE PROBLEMS

See "Appendix B: Answers to Quizzes, Chapter 9"

(Use appendix B to grade your quizzes by assigning two point for each totally correct answer, and one point for each partially correct answer. The answers do not need to be verbatim but should convey the correct concepts. If the score is below 85%, restudy the chapter and retake the quiz; If need be repeat this process until the score is above 85% and only then continue to the next step).

 **SUMMARY**

Lesson 9- Summary
The following important points are the highlights of this lesson:

1. Waves form the final piece of the puzzle in the emerging science of electricity, from electrostatics and electrokinetics to electrodynamics and electromagnetics.

2. A physical model for the electromagnetic force fields delivered two important discoveries: a) displacement current, and b) EM waves, which propagate at the speed of light.

3. Maxwell summarized his work into eight and then into four concise equations, which clearly and exactly expressed all of his findings.

4. The concept of ether was discarded as a false datum in 1887 by an American scientist named A. A. Michelson, as a nonexistent medium.

LESSON 10

The Experimental Proof of Waves

 **LEARNING OBJECTIVES**

After completing this lesson, you will be able to:

1 Understand the methods that Hertz used to verify the validity of the EM theory as proposed by Maxwell.

2 Understand what the scientific climate and the conflicting views concerning electricity were before the dawn of truth.

3 Understand what the exact properties of waves are.

4 Recognize how the discovery of electrons and photons changed the way we view electricity and light today.

5 Understand how Tesla's contribution is still being used in the power distribution systems toady.

READING ASSIGNMENT

Read the following sections of Chapter 10 of the textbook:
Sections: 10.1-10.11.
These sections discuss the basic concepts of how the validity of waves was verified experimentally.

KEY POINTS

1. Hertz showed experimentally that:
a) EM waves, though invisible, actually exist physically, and
b) Light waves are of electromagnetic origin and obey all of the electromagnetic laws as set forth by the Maxwell's equations.

2. Based on light wave characteristics, four properties of EM waves were investigated and demonstrated to exist:
a) Reflection,
b) Rectilinear propagation,
c) Polarization, and
d) Refraction.

3. The discovery of electrons and photons verified the duality of waves and particles for electricity.

4. Tesla's use of a transformer and AC power distribution system eliminated the DC power systems and laid the groundwork for modern power systems.

PRACTICE PROBLEMS

Do Quizzes 10.1 and 10.2.
(This is a self test and one must use the "Honor Code" while taking this test).

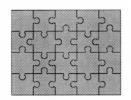

ANSWERS TO PRACTICE PROBLEMS

See "Appendix B: Answers to Quizzes, Chapter 10"
(Use appendix B to grade your quizzes by assigning two point for each totally correct answer, and one point for each partially correct answer. The answers do not need to be verbatim but should convey the correct concepts. If the score is below 85%, restudy the chapter and retake the quiz; If need be repeat this process until the score is above 85% and only then continue to the next step).

 **SUMMARY**

Lesson 10- Summary
The following important points are the highlights of this lesson:
1. The first thing Hertz did was to establish the existence of EM waves.

2. The four properties of waves that were verified experimentally next, were: a) reflection, b) propagation, c) polarization, and d) refraction.

3. Beyond a shadow of doubt, Hertz showed that light is also of EM origin and thus made it a subset of electricity.

4. Hertz showed how to produce high-frequency waves using the concept of displacement current proposed by Maxwell.

5. Tesla's pragmatic ideas using the AC system beat out Edison's DC system and won the world over in the power generation, transmission, and distribution arena.

LESSON 11

The Supplemental Discoveries

 LEARNING OBJECTIVES

After completing this lesson, you will be able to:

1 Understand what the supplemental discoveries are.

2 Realize that there are eight supplemental discoveries that aid and abet the six monumental discoveries in bringing about a unified theory of electromagnetics.

3 Understand the electronic structure of materials.

4 Understand the conditions under which a material can achieve superconductivity.

5 Realize that every supplemental discovery is accompanied by a magnetic dual which makes up a complete set.

6 Understand that a) the principle of conservation of charge, and b) the principle of charge migration to a conductor's surface, are both byproducts of the supplemental discoveries.

READING ASSIGNMENT

Review the following sections of Chapter 11 of the textbook: Sections 11.1- 11.11.

These sections discuss the supplemental concepts that fill in the gaps and augment the major discoveries of electricity.

KEY POINTS

1. The eight supplemental discoveries are accompanied by their magnetic duals and facilitate the understanding of the monumental discoveries and create simplifications in practical problems.

2. The First supplemental discovery: charge is the source of E-field. Its magnetic dual states that there is no magnetic charge.

3. The second supplemental discovery is the linearity of current and voltage for DC signals. Its magnetic dual states the linearity between the magnetic flux and the magnetomotive force.

4. The third supplemental discovery states the linearity between AC current and voltage in the phasor domain. There is no magnetic dual.

5. The fourth supplemental discovery states that there is self induction when there are changes in the current in a circuit.

6. The fifth supplemental discovery is about the absence of absolute charge from which two corollaries can be derived:
5a) Electric charge is the result of separation of opposite charges existing in a neutral particle, and
5b) Electric charge always migrates to the surface of a perfect conductor regardless of its initial location.

7. The sixth supplemental discovery concerns the existence of electric polarization, and its magnetic dual, which is the magnetic polarization.

8. The seventh supplemental discovery is about the conservation of current with its magnetic dual stating that magnetic flux is also conserved.

9. The eight supplemental discovery states the conservation of voltage with its magnetic dual stating the conservation of magnetomotive force.

PRACTICE PROBLEMS

Do Quizzes 11.1 through 11.7.
(This is a self test and one must use the "Honor Code" while taking this test).

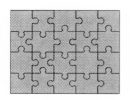

ANSWERS TO PRACTICE PROBLEMS

See "Appendix B: Answers to Quizzes, Chapter 11"

(Use appendix B to grade your quizzes by assigning two point for each totally correct answer, and one point for each partially correct answer. The answers do not need to be verbatim but should convey the correct concepts. If the score is below 85%, restudy the chapter and retake the quiz; If need be repeat this process until the score is above 85% and only then continue to the next step).

SUMMARY

Lesson 11- Summary
The following important points are the highlights of this lesson:

1. The supplemental discoveries provide a series of simplified truth about the field of electricity and magnetism, and help us to provide solutions to complex problems.

2. From the eight supplemental discoveries a number of laws and corollaries can be derived, which help establish a pool of practical knowledge for easy analysis or design.

3. A much deeper understanding of the phenomena of electromagnetism can be gained by knowing the supplemental discoveries and their magnetic duals than merely grasping the concepts behind the monumental discoveries.

4. Many subjects, such as electric circuits, electronics, magnetostatics and so on, can be easily understood and applied with tremendous success from a study of the supplemental discoveries and their magnetic duals.

LESSON 12

Beyond the Modern Age

 LEARNING OBJECTIVES

After completing this lesson, you will be able to:

1 Understand the role of postulates in sciences.

2 Understand the exact sequence of postulates used in the construction of the physical universe.

3 Know the three categories of application mass and how they apply to one's life.

4 Understand the applications of Nano-technology.

5 Know what lies in the universe ahead.

6 Understand that purposes lie beyond postulates and therefore are senior to postulates.

READING ASSIGNMENT

Review the following sections of Chapter 12 of the textbook, sections: 12.1-12.16.
These sections discuss the concepts and factors that contribute to the current condition of modern age and what lies beyond it.

KEY POINTS

1. Our universe is based on a series of exact postulates, which sciences have unearthed and discovered to be the common denominator of any and all physical entities, objects, forms, etc.

2. The concept of application mass can be generalized to include the materials not produced by Man, which instantly takes us either into the microscopic world or into the unlimited space of heavens and thus presents us with a before-hand knowledge base about the unknowns in our universe, long before their discovery in a very distant future.

3. The validity of the big bang theory is addressed and the fact that it is limited in scope of analysis,

renders it useless for long-range understanding of our universe.

4. There is a relationship between the mechanics of space, energy, matter and time and the universe of thought and the field of considerations in that the former is the direct byproduct of the latter and the connecting string between the two is postulates. The postulates through a series of exact bands gradually trickle down and finally solidify or materialize into the mechanics of space, solids, objects and the whole panoramic material universe.

5. Application mass can be categorized as:
a) Generalized application mass such as rocks, planets, rivers, etc.
b) Technical application mass such as cars, computers, TVs, etc., and
c) Personalized application mass such as art forms, personal jewelry, etc.

6. The science of physics is based upon a collective viewpoint through centuries and has not adequately answered the question of why classical and quantum physics are so drastically different.

7. The science of physics and engineering have studied the effect side of life and have completely ignored the cause side of life and what has actually caused our universe to come about.

8. The emerging science of nanotechnology will have a huge impact on our future in a technologically

oriented society as well as the way we will manufacture technical application mass.

9. The physical universe is seemingly an infinite universe with vast distances between planets and galaxies, which makes traveling to its edges an impossibility. However, this seeming reality can be understood better if we could achieve an exterior viewpoint purely as an imaginary exercise on a mental level, at which moment we could suddenly realize its finite size.

10. Beyond postulates, there is a much senior concept: purposes, which dictate certain postulates to come about and be implemented. Understanding the main purposes leading to our current universe's postulates and materials is beyond the scope of this work at the moment.

 **PRACTICE PROBLEMS**

Do Quizzes 12.1 through 12.6.
(This is a self test and one must use the "Honor Code" while taking this test).

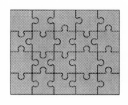

ANSWERS TO PRACTICE PROBLEMS

See "Appendix B: Answers to Quizzes, Chapter 12"

(Use appendix B to grade your quizzes by assigning two point for each totally correct answer, and one point for each partially correct answer. The answers do not need to be verbatim but should convey the correct concepts. If the score is below 85%, restudy the chapter and retake the quiz; If need be repeat this process until the score is above 85% and only then continue to the next step).

SUMMARY

Lesson 12- Summary
The following important points are the highlights of this lesson:

1. The role of postulates in sciences is essential and can never be discounted under any conditions.

2. Application mass is the visible end of the invisible postulates at the mental level.

3. Nano-technology is based upon quantum mechanical concepts and will play an important role in the direction of our future technology.

4. The universe ahead is a world for sciences and promises to be a fruitful ground for discoveries concerning the exact connection of life force and/or the mental phenomena to the postulates, which have led to the ever-dominant material universe.

LESSON 13

Conclusion

 LEARNING OBJECTIVES

After completing this lesson, you will be able to:

1 Know how considerations play a primary role in all aspects of the physical universe.

2 Understand how every piece of matter is based on the created space, created energy, created mass, and created time side of existence.

3 Realize how and why matter is a dense standing wave.

4 Know how the microscopic and macroscopic universes intertwine and give us a shared reality.

5 Realize that the uncertainty principle is a built-in feature of the material universe and we are dealing with it on many levels.

6 Understand the mindset of God and know what the thought pattern was when the physical universe was being created at the "blueprint" stage.

READING ASSIGNMENT

Review the Chapter 13 of the textbook, sections 13.1-13.5. This chapter discusses a summary of the fundamental concepts underlying our physical universe.

KEY POINTS

1. The imposing and ever-dominant physical universe is based on a series of considerations, which seem to be unalterable to the average man.

2. The certainty that we observe on a classical level of observation is an apparency based on the uncertainty of particles at the quantum level.

3. The millions and billions of considerations concerning the physical universe have a common denominator of four postulates, which can be further reduced to three exact postulates. This is a major simplification of our complex universe which has enormous ramifications in any science.

4. Knowing the mind of God means knowing the subject of postulates and recognizing their power in

bringing about any universe in general, or any application mass in particular.

5. The microscopic universe is the actuality version or the underlying theme of our universe whereas the macroscopic universe is the apparency version or the visible version of the same universe but viewed from a different perspective.

 **PRACTICE PROBLEMS**

Do Quizzes 13.1 through 13.4.
(This is a self test and one must use the "Honor Code" while taking this test).

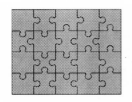 **ANSWERS TO PRACTICE PROBLEMS**

See "Appendix B: Answers to Quizzes, Chapter 13"

(Use appendix B to grade your quizzes by assigning two point for each totally correct answer, and one point for each partially correct answer. The answers do not need to be verbatim but should convey the correct concepts. If the score is below 85%, restudy the chapter and retake the quiz; If need be repeat this process until the score is above 85% and only then continue to the next step).

SUMMARY

Lesson 13- Summary
The following important points are the highlights of this lesson:

1. The physical universe is an average universe based on the same very principle that any other universe is made from except that it appears to be a much more imposing and dominant universe, and extremely commanding once one agrees with it and gets placed in it.

2. The postulates making up the physical universe were conceived trillions of years ago, but at each moment in time one needs to regenerate them in order to perceive any portion of the universe at all.

3. The energy in the physical universe is in constant agitation and gives rise to time.

4. Matter is a dense standing wave, which is floating in time and appears to be relatively unaffected by the passage of time. The higher density of matter indicates that it endures longer and thus is more timeless. An example of a high density material is diamond and the old maxim "Diamonds are forever", takes its cue from this fact.

Appendix B

ANSWERS TO QUIZZES

CHAPTER 1
Answers

QUIZ 1.1

1.1 What are the mechanics of creation of any universe?
Steps a-f in section 1.3.2:

a) *VIEWPOINT – The first step is the assumption of a viewpoint. This is the first and foremost step in the creation of any universe.*

b) *SPACE – Have the viewpoint set up a region, which is delineated by a boundary surface. This region effectively creates a workable space in which all future creation can take place.*

c) *PRIMARY POSTULATES – Put forth a series of primary postulates, unconditionally true at all times, which are capable of creating different objects, each with its own distinctive characteristics and peculiarities. The term object, being the product of a postulate, should be taken conceptually and not literally, since in the field of arithmetic, for example, there are exact postulates, which have created "10 objects", numbers 0 through 9.*

d) *WORKABLE & AUXILIARY POSTULATES – Put forth a series of workable and auxiliary postulates, which define how the created objects will interact with each other. For example, in the field of arithmetic, there are four workable postulates, which create "four operations", which are addition, subtraction, multiplication, and division.*

e) **LAWS, RULES & THEOREMS** – *Obtain a series of laws, rules and theorems from these postulates, which will uniquely define the interactions and inter-relations amongst the created objects. For example, in the field of arithmetic, the "commutative law" can be obtained purely by observation: 1+2=2+1. Therefore, steps (a-e) define all the considerations built into the field of arithmetic.*

f) **APPLICATION MASS** – *The Final aspect of any postulated universe is the creation of its application mass, which approaches infinity in sheer number. For example, in the field of arithmetic, the application mass includes calculators, computers, computer codes, computer software and programs, ledgers, accounting balance sheets, abacus, etc.*

1.2 What is meant by a scientific postulate?
It is a scientific assumption that requires no proof and is assumed to be true unconditionally and for all times, used as a basis for reasoning.

1.3 What is a scientific axiom?
Is a self evident truth accepted without proof.

1.4 Why is nomenclature so important in understanding a science?
Because it packages concepts into specific words, which form the woof and warp of the scientific language,

essential for the communication and understanding process.

1.5 What role philosophy plays in a science?

Sciences have greatly benefited from the field of philosophy in that they have adopted a practical philosophy known as "scientific methodology," which requires experimental proof of any proposed theory, thus aligning well with "what is" rather than "what is theorized to be." This is the exact difference between physics and psychology.

1.6 What is the role of mathematics and mathematical models in sciences and engineering?

Mathematics are short-hand methods of stating, analyzing, or resolving real or abstract problems and expressing their solutions by symbolizing data, decisions, conclusions, and assumptions.

1.7 What is scientific methodology? Why is it superior to other methods?

Scientific methodology as given in section 1.8.3, is a systematic approach based on a) observation,   b) collection of data, d) classification of data, e) formulation of a hypothesis, e) testing and experimentation, f) repeat if refinement needed and g) emergence of a workable theory.

It is extremely superior to other methods because it delivers a workable answer that can be used over and over to solve any new problem and help us in predicting the future behavior of any complex system.

QUIZ 1.2

1.8 What has made the physical sciences superior relative to social sciences?

The identification of exact postulates along with axioms, theorems and scientific methodology has produced a solid background of technical data against which any problem can be resolved with technical accuracy. Social sciences have not done this but have wandered into the field of hypothetical propositions without the use of scientific methodology. As a result, their pseudo-sciences have very limited workability and can not be put to acid test of proof.

1.9 What is Einstein's theory of Relativity?

It is a theory about motion and moving reference frames and how different reference frames may experience different measurement results for the observed time and speed as well as accuracy of results.

1.10 What is meant by special theory of relativity? What is the general theory of relativity?

Special theory of relativity proposes that all natural laws are the same for observers in frames of reference moving with constant relative speed. The general theory of relativity generalizes the special theory of relativity to nonlinear and non-inertial frames of reference, where gravitation is incorporated and in which events take place in a curved or nonlinear space.

1.11 What is the principle of relativity of knowledge?

Any datum derivable from a postulate or axiom can never have an absolute value but only a relative value dependent on the viewpoint

1.12 What is the pyramid of knowledge?

It is a hierarchical approach to classifying information so that the most important and senior data is closer to the top.

1.13 What is meant by the duality principle?

It is a principle that for any theorem, there is a dual theorem in which all quantities and operations are replaced by their duals.

1.14 What is meant by the dichotomy of static and kinetic?

Static (non-moving) as cause point is the dichotomy of kinetic (moving), which is at effect point and thus junior to static. In other words, before there was kinetic electricity, for example, there was static electricity.

QUIZ 1.3

1.18 Why is practical knowledge hierarchical, in the form of a pyramid?

Because all data emanate from a viewpoint at the apex, and proceed through postulates, level by level down to a plethora of application mass at the base of the pyramid.

1.19 What is meant by evaluation of information? What are the steps required to evaluate any datum?

The value of information must be established first before one can use it effectively. The process of evaluation of information is established by the six steps delineated in section 1.4.4 as follows:

The Evaluation process

 1. *Discover the underlying postulates of the science.*

2. *Establish a series of laws, facts and known information, directly derived from the postulates. This series of laws and facts would serve as the basis for the formation of any desired subject under study and will establish the "touchstone" essential to any evaluation process.*

3. *Compare the new datum to the series of postulates, laws and facts established in (1) and (2).*

4. *Determine the degree of alignment and understanding it provides to the known and existing facts.*

5. *If the degree of alignment is high, it should be assigned a high value.*

6. *If, on the other hand, it is an arbitrary factor or is irrelevant and does not align and explain the existing and known facts, then it should be assigned a low or no value and discarded from further or future utilization.*

1.20 What are the necessary steps that are required to solve any problem?

To solve any technical problem, one needs to follow the steps delineated in section 1.8.1 as follows:

Step #1	*Dissect the problem into sections of similar and related data.*
Step #2	*Compare each area to already known natural laws.*
Step #3	*Use the natural laws, obtain a solution for each section.*
Step #4	*Combine and integrate the solutions so obtained (from step #3) into one comprehensive solution called the*

First-Order Solution (Also called the Baseline Solution)

Step #5   *Resolve the remainder, which can not be known immediately, by using the known part (and its solution) to arrive at the second-order solution.*

Step #6   *Refine the second order solution if the desired degree of accuracy is not reached. The process of refinement has to be repeated "n-times", if needed, to bring about a more accurate solution, referred to as the Final Solution, or the "nth-Order Solution".*

1.21 What is meant by false information? And how one can avoid it?

By false information we mean information which is contrary to known laws and axioms and yet may seem to be true by an uneducated observer. For example, the datum "war is good for economy" may be the stock in trade of some politician who completely believes in its validity, but happens to be a totally false data, based on the politician's misunderstandings.

We can avoid it by knowing the basic laws and axioms of the subject so well that we can easily detect any advertised or proposed false data.

1.22 What does the workability of a scientific postulate depend upon?

The degree of workability of a postulate is established by three factors:

a) The degree it explains an existing phenomenon,

b)   The degree it predicts new phenomena hitherto unknown, and

c) The degree that it does not require arbitrary data to be called into existence for its explanation.

The above three factors form a powerful set of criteria, against which any scientific premise or, in general, any proposed data can be tested for its potential worth with tremendous success.

1.23 What is the definition of a physical science?

A Physical Science is the study and analysis of the inanimate components of the physical universe, excluding the life force, with the intent of using the principles and results thus found for the betterment (or otherwise) of the conditions of mankind or all life organisms that are symbionts to existence. By symbiont we mean an organism intimately living with another, especially when in a state of mutual advantage (e.g. a pet, etc.).

1.24 What is the mental framework of an average scientist in relation to modern sciences and their associated postulates?

The average scientist's awareness is well below the level of existence of postulates and their influence as the main driving force behind our universe. This has been the primary factor, which has necessitated the author to make this commentary, and to point out what is propelling the physical universe into existence—it is not the raw force but the postulates behind it! This is a rather gross oversight by the scientific community. In other words, for a scientist to mistake the application mass for actuality and totally neglect the postulate as the main proponent of its existence is rather unforgivable, now that we know the power of postulates in sciences.

In other words, the cause-effect relation of the physical universe has been lost and the sciences have been turned into the study of effects. In other words modern sciences have taken on the study of created spaces, created energies, created masses and the resultant created time periods. Our "dyed-in-the wool scientist" has taken the role of the observer and reporter of effects.

1.25 How could the subject of viewpoint revolutionize the way sciences are studied?

The subject of viewpoint is a new approach to understanding scientific postulates, concepts, laws and considerations. This approach not only simplifies, but generalizes the subject of physics and engineering thoroughly. It adds meanings to the basics where there were none.

1.26 What does this work intend to achieve and what is its most important goal, as set forth by the author?

With the help of this work, the author intends to:

I.   *Create a better appreciation of physical sciences in general, and change the way sciences are perceived, more of the order of "precious information", which is vital to the survival of human species,*

II. *Promote better and deeper understanding of the scientific methodology, and how it is used to achieve stellar results in the scientific arena,*

III. *Bring about a public awareness of frailties involved with social sciences and how to align and improve these vital subjects better, and*

IV. Finally and most importantly, change the course of physical sciences not only in the way it is taught across the land at high schools, colleges, universities, etc., but more significantly alter the methods of their applications and steer them toward more humane and ethical causes.

CHAPTER 2
Answers

QUIZ 2.1

2.1 Where is man's home relative to all other stars in our galaxy?

In this vastness of space, high speed energy particles and enormous number of masses of all shapes and sizes, which have persisted over many trillions of years, we find earth a small planet on a remote solar system. This is where man's home is as he wheels through the physical universe. Man's short time span of existence as a species (between 35,000 and 100,000 years) and his current state of survival on earth, is an interesting situation, which will be the subject of much ponder for many years to come.

2.2 What is meant by the shared universe?

We are dealing with an enormous mass of bodies and vast spaces between them. Nevertheless this is the universe, through a small portion of which, we communicate to others or exist on a physical plane. Since the physical universe is shared by every being, it is referred to as the shared universe.

2.3 What is meant by the primary postulates of the physical universe?

In the quest for knowledge of the blueprint of our physical universe we need to have a basic understanding of its constituent postulates, which consequently brought forth the aggregate of masses and energy forms in vast spaces of unfathomable extent. Thus to study the physical universe, we need to examine some very fundamental materials, which are of utmost significance to a thorough knowledge of the subject. These fundamental materials we refer to as primary postulates.

2.4 What is the first primary postulate and what is its significance?

The study of the existence of matter and energy as done in physics, presupposes the existence of a linear created space in which we are placing these entities in. Linear space is the first consideration behind the science of physics, which is implicitly alluded to but not directly addressed.

Its significance is that it is the first and foremost point of establishing any physical entity.

2.5 What is a reference frame and why do we need it?

Reference frames are essentially an organizing tool set up to measure a created space. Reference frames organize the created space and through the measurement system that they employ, one may consider them a method of bookkeeping of that space. Their main functions are:

a) To provide orientation,

b) To designate a reference point, and

c) To identify each point uniquely in the three dimensional created space.

2.6 What is meant by curved space? Can space actually be curved? Explain.

Created space is the delineated and designated region within a closed boundary line (2-dimensional space) or a closed and bounded surface (3- dimensional space), within which all things under consideration can be placed.

It should be noted that under heavy force fields (such as strong gravitational fields, electric or magnetic fields), space can appear to be curvilinear, which means that the shortest path between two points is along a curved line and not a straight line. The notion of curved or nonlinear space is an apparency and not an actuality. The actuality is that the strong fields bend the flow lines and give the illusion of nonlinear space. However, mechanical space remains linear.

2.7 What is a space warp and does it exist? Explain.

The concept of space warp, wherein the created space has a discontinuity in it or acts nonlinearly, can safely be considered as purely a theoretical concept at best. Space warp is an abstract concept, which exists in the thought universe. However, as a concept it can be easily expressed mathematically and thus one can work out all of its properties in terms of the characteristics of the warp.

QUIZ 2.2

2.8 What is the second primary postulate of the physical universe?

The Second postulate deals with the concept of Energy— the universal force that the ancient philosophers have talked about. The concept of energy in the physical universe takes on many forms. Energy is encountered in all aspects of existence, particularly in the scientific arena. It is one of the foremost entities, which deserves an exact definition:

DEFINITION- ENERGY: *is defined to be the capacity or ability of a body to perform work. This definition can be applied to particles and objects in both quantum and classical physics. Furthermore, energy of a particle or object can be subdivided into two parts:*
a) Potential energy due to its potential motion, and
b) Kinetic energy due to its actual motion.

2.9 What is the difference between quantum physics and classical physics?
In simple terms, classical physics deals with large objects and waves following the Newtonian mechanics and Maxwell's equations, whereas quantum mechanics deals with small particles (e.g. atoms, electrons, etc.), which are the constituents and sources of the large objects and waves.

2.10 What are the two classifications of motion?
Macroscopic or large motion and microscopic or small motion.

2.11 What are the subdivisions of large motion?
Flow, divergence, and standing wave.

2.12 What is the third primary postulate of the physical universe?
The third Postulate is matter. Matter and energy were considered separate entities for a long time until recent discoveries and advances, which made them equivalent.

Thus based on this understanding we need to define matter at this point.

DEFINITION: MATTER- *Matter particles are the result of bringing energy particles into close proximity where they occupy a very small volume.*

2.13 What are the two subdivisions or types of time? Explain each.

Mechanical time and subjective time as follows:

MECHANICAL TIME *is defined as a measurable extent of change--change of "position of a particle" relative to a starting point in a created space. It is a primary postulate used in the creation of the physical universe (on a macroscopic and/or microscopic level) and its value at any given location orders the sequence of events at that point in space.*

SUBJECTIVE TIME *is defined as the consideration of time in one's mind, which could be called the subjective time, and is a nonlinear or linear quantity depending on one's viewpoint.*

2.14 What is the difference between objective and subjective time?

The objective time can be measured in the physical universe very accurately within a fraction of a second. Each segment of it can have a beginning and an exact ending point. On the other hand, the subjective time is one continuous stream of change with no specific beginning or end and only gets a time stamp from the objective time side of existence.

2.15 Can time be referred to as the fourth dimension? Why?

There is a considerable discussion on time as a fourth dimension amongst physicists. They see the change taking place in a three dimensional created space of the physical universe and they want to plot this change as a fourth dimension of the physical universe.

There are several technical problems associated with this action. First of all, time and space are not cut from the same cloth, so to speak. Space is senior to energy and matter. Since time is a result of their interaction—a derived and secondary byproduct of this interaction, therefore it is junior in importance relative to space. Thus space and time are not of comparable magnitudes!

Secondly, Time in the strictest sense of the word is a postulate about change (see stratum 1 in Figure 2.14). However, actual change in the physical universe as represented as time in physics texts is the second stratum, which is a cut-down from this postulate. One should recall, before there was change there was the postulate of time. Therefore the two strata are not the same thing.

Also, as a counter argument, we can see that the concept of fourth dimension is dependent on the change of particle's position and interaction of matter at each point in space. However, if there are no matter and energy to interact at that point in space, then there is no change thus the fourth dimension instantly disappears! In other words, space can exist without any matter or energy in it thus it would have no time associated with it and it would be timeless. Therefore, time could possibly be considered a fourth pseudo-dimension, at best.

QUIZ 2.3

2.16 What are the original postulates?

The three original postulates can be summarized as:

I. Space

II. Energy (or Force)

III. Change

Combining postulates (II) and (III), we can actually obtain the third and the fourth primary postulates discussed earlier (i.e., matter and time). For example, if we "change" the created space, in which certain energy particles exist, in such a way that the volume is reduced, then the energy particles become more condensed and thus we get "matter" (the third primary postulate).

2.17 What is the missing vital element?

Viewpoint.

2.18 What is meant by the concept of viewpoint and why is it important?

Before the three original postulates or the four primary postulate came into existence, postulate zero existed. We define this postulate now:

POSTULATE ZERO: *Is the existence of viewpoint as the primary factor for all subsequent byproducts and actions. One of the main abilities of a viewpoint is the ability to assume beingness. It is an important point to understand because it leads to resolution of many tangled mysteries hitherto unresolved.*

2.19 What is meant by "created space" and "creating space"? Why are they a pair in dichotomy?

The concept of "created space" is all what physics has currently concerned itself with and has actually made it synonymous with "space". However, from the dichotomy principle, we know that there must exist an equal but contrary datum. The dual of "created space" is "creating space." To create space on a conscious level and not on an automatic basis requires a viewpoint. As we discussed in Chapter 1 (Section 1.3.2 entitled "The Mechanics of Creation of any Universe") the first action will be to set up a region delineated by a boundary surface. Such an action instantly creates space. For example, designing a circuit or an automobile requires one to decide on the space or size of the object before one can build it. Such an action could be classified as "creating space".

2.20 What is the generalized concept of duality?

From a general observation of life and livingness, we can see that the principle of dichotomy can be applied to any of the three remaining components of the physical universe (energy, mass and time). Take for example energy. We have "created energy" of the sunlight, which is constantly and automatically created by the sun on a regular basis. The dichotomy of "created energy" would be "creating energy". An example of creating energy would be composing a fast paced music, which would cause an audience to create a high-energy level of action and excitement.

Similarly, the dichotomy of "created mass" is "creating mass". An example of created mass is an atom. It has been created and put there for observation, study or use. On the other hand we have creating mass, which would be combining two atoms to create a new atom (fusion) or splitting an atom into simpler atoms (fission). These two could be classified under the heading of creating mass, since the new atoms thus created are surely different than the original mass.

The principle of dichotomy can be generalized to "time" just as well. The dichotomy of the "created time" is "creating time". The viewpoint can create time by shifting one's viewpoint and create a whole new time stream completely separate and independent of the mechanical time.

2.21 Why "design" and "analysis", in general, are dual of each other?

"Engineering analysis" of a desired object is dealing with and taking apart its created space, created energy, created mass and created time stream. On the other hand, "Engineering design" brings the dynamic factor of "creation of space, energy, mass and time" into focus. Therefore, we can observe that from the point of view of engineering, "analysis and design" are concepts in dichotomy.

QUIZ 2.4

2.22 What is an axiom?
A self-evident truth accepted without proof.

2.23 What are the three physical universe axioms?
a) Uniqueness axiom,
b) Mutual-exclusiveness axiom, and
c) Non-absoluteness axiom.

2.24 What is the uniqueness axiom?
THE UNIQUENESS AXIOM: *Given an exact set of necessary and sufficient initial or boundary conditions,*

any problem in the physical universe (excluding design)
has a unique and exact solution.

2.25 Why social sciences have not developed the "unique-solution type" status?

The important point to remember here is that the
Uniqueness axiom only applies when the problem under
consideration exists in the "physical universe" with all
necessary assumptions exactly specified. This is unlike
the social sciences where there are no solid
fundamentals and as a result many arbitrary solutions
could exist to one problem where all seem to provide a
seemingly satisfactory answer to a social problem, but
none of the solutions are unique (i.e., all are arbitrary
solutions!).

2.26 What is the Mutual-exclusiveness axiom and its application to sciences?

THE MUTUAL-EXCLUSIVENESS AXIOM: *Two physical*
bodies or particles of any size can never occupy the
same space.
The direct result of this axiom is that every created
space does not overlap any other created space.   This
axiom clearly establishes the concept of separateness or
non-overlapness of the physical universe and forms the
foundation upon which each created space continues to
remain stretched out and uncollapsed by all other
created spaces.

2.27 What is the Non-absoluteness axiom and how can we use it in sciences?

THE NON-ABSOLUTENESS AXIOM: *This axiom states that*
any and all measured quantities, derived values,

scientific laws, observed principles and technical data connected with the physical universe can not be expressed in absolute terms, rather in relative, limited, and highly qualified statements.

This axiom is a direct consequence of Heisenberg uncertainty principle, which is interwoven into any and all particles of matter and their properties, whether of microscopic or macroscopic size.

2.28 What do we mean by "physical universe" as a generalized application mass?

Through the use of inductive logic, we can generalize the concept of man-made application mass to other natural forms of application mass such as planets, stars, etc. This may be at first a relatively startling conclusion and a stretch of logic but a "natural" or "man-made" application mass follow the same pattern of development as delineated in the pyramid of knowledge. Therefore, the natural application mass, which are not a direct byproduct of a science, is called the "generalized application mass", in this work.

2.29 What is meant by postulate and application mass as being a subset of "cause–effect" pair?

The "Cause-Effect" concept of "Postulate-application mass" can be understood more clearly if we delineate the properties of each and the draw our own conclusions.

A Postulate, as the beginning point, of any creation can be considered to possess the following properties:

a) Zero space,

b) Zero mass,

c) Zero time dependency i.e. is timeless,

d) Zero energy content,

e) An invisible quality.

An application mass (or a generalized application mass), as the end point, of any creation can be considered to possess the following properties:

a) Finite space,

b) Finite mass,

c) Finite lifetime,

d) Finite energy content,

e) A visible quantity.

2.30 Why is space senior to all other components of the physical universe? Give an example.

Because space has to exist first before any physical entity such as matter, energy or time can be conceived to exist. For example, in choosing a suitable car the primary consideration is "what size car do I want?" This concerns its shape or the space it occupies. Then one considers its power color or other factors.

CHAPTER 3
Answers

QUIZ 3.1

3.1 What are the two interwoven universes of physics?
Physics is divided into two fields:
a)   Classical Physics, which itself is further subdivided into electricity, magnetism, optics, acoustics, mechanics and heat; and
b)   Quantum Physics, which is subdivided into atomic physics, nuclear physics, and solid-state physics.

3.2 What are the two workable postulates of physics? Define each and give an example.
WORKABLE POSTULATE #1*: Let energy exist in fragmented forms, which leads to existence of potential energy, kinetic energy, work, power, force, field, voltage, current, pressure, and wave.*
WORKABLE POSTULATE #2*: Is the existence of matter in the form of particles (microscopic), or grouped particles called objects and bodies (macroscopic).*

3.3 What are the two subdivisions of energy?
*We can observe that the subject of energy deals with either **"potential" or "actual" motion**. Thus energy could be subdivided into two categories:*
a) Potential energy and
b) Kinetic energy

3.4 What are the electrical forms of potential energy?

Electrical forms of potential energy include:
1) Static charge
2) Static force
3) Static field
4) Voltage (or electrical potential difference)
5) Magnetic Pole
6) Pressure

3.5 What are the electrical forms of kinetic energy?
Electrical forms of kinetic energy include:
7) Momentum (linear and angular)
8) Time-varying or moving electric charge (or Current)
9) Dynamic Force
10) Dynamic field (Electric and Magnetic)
11) Power
12) Electromagnetic waves

3.6 What are the two implicit postulates of physics?
IMPLIED POSTULATE #1: *Is the existence of a continuous and linear "created space".*

IMPLIED POSTULATE #2: *Is the existence of a continuous and linear index of motion commonly known as mechanical time, measured in seconds, minutes, hours, etc.*

QUIZ 3.2
3.7 What is the most vital ingredient in physics and engineering?
It is a viewpoint defined as "a point, on a mental plane, from which things are perceived and judged".

3.8 What are the two main subdivisions of viewpoint?

POSTULATING VIEWPOINT: *is a point in the thought universe, which is potentially capable of creating a postulate, a field of study, any and all components of the physical universe, a physical object or in general, a complete universe.*

OBSERVING VIEWPOINT: *is a point in the thought universe, which without any alteration, is capable of pure observation of an idea, a postulate, a field of study, any and all components of the physical universe, a physical object or a whole universe.*

3.9 Why is a coordinate system a poor substitute for a viewpoint? Explain.

A reference point substitute for a viewpoint is poor because a reference point in a coordinate system is purely a physical point in space and not truly a viewpoint in the full sense of the word.

3.10 What are orthogonal coordinate systems?

Orthogonal coordinate systems are comprised of three families of surfaces that are mutually perpendicular to each other and each physical point in space can uniquely be specified by the intersection of the three exact surfaces.

3.11 What are the three types of coordinate systems used today in sciences? Describe and draw a diagram for each type.

a) Cartesian (or Rectangular Cartesian or simply Rectangular) Coordinate System
b) Cylindrical Coordinate System
c) Spherical Coordinate System (also known as Spherical Polar Coordinate System)

QUIZ 3.3

3.12 What is the first hidden postulate of physics? Explain in simple terms what it says and give an example.

HIDDEN POSTULATE #1: *Is the principle of conservation of energy, which is held valid under all conditions and for all times.*

Trying to heat a gallon of water would require us to use chemical energy (burning coal) to generate electricity which would then be converted to heat to warm up the water.

3.13 What is the second hidden postulate of physics? Give an example.

HIDDEN POSTULATE #2: *Is the principle of conservation of mass, which is held valid under all conditions and for all times.*

As an example, if we mutilate a form of matter by burning it, the mass remains the same provided no mass to energy conversion process (such as in radio active material) took place.

3.14 What is the third hidden postulate of physics? Give an example.

HIDDEN POSTULATE #3: *Is the principle of conservation of energy-mass, which is held valid under all conditions and for all times.*

An example could be a nuclear explosion, where part of the mass converts to energy.

3.15 What is the fourth hidden postulate of physics? What is an exception to this postulate?

HIDDEN POSTULATE #4: *Is the principle of conservation of created space, which is held valid under all conditions and for all times.*

An example could be putting two created spaces together, such as a TV and a computer together, which is the addition of the two spaces.

An exception is when the created space does not have a rigid boundary such as a piece of soft plastic, an open box, etc.

3.16 What is the fifth hidden postulate of physics? Explain in simple terms what it means.

HIDDEN POSTULATE #5: *Is the principle of conservation of mechanical time, which is held valid under all conditions and for all times. An example could be watching two movies each two hours, requires a four hour period of ones life.*

3.17 What is the sixth hidden postulate of physics? Explain its implications in scientific laws.

HIDDEN POSTULATE #6: *Is the principle of comparable magnitudes, which is held valid under all conditions and for all times. An example would be adding two numbers of the comparable magnitudes such as 2+5, unlike 0.001+1000.*

3.18 What is the principle of bilateral communication? Explain its relation to the Newton's law of interaction.

THE PRINCIPLE OF BILATERAL COMMUNICATION: *Once a communication channel has been established between any two physical entities of comparable nature and size, then that communication channel is always bilateral and* *inevitably affects both entities. In other words, creation of a unilateral physical communication channel is impossible!*

Newton's law of action and reaction is a byproduct of this powerful principle.

QUIZ 3.4

3.19 What is meant by a wave and a particle?

They are opposite of each other since they are defined as:

DEFINITION-PARTICLE: *Any infinitesimal subdivision of matter, which is ranging in diameter from a fraction of an Angstrom (such as an electron, an atom or a molecule, etc.) to a few millimeters (such as a raindrop, etc.). This concept is shown in Figure 3.13.*

DEFINITION-WAVE: *A disturbance or vibration which propagates from one point in a medium to other points without giving the medium as a whole any permanent displacement. Waves follow certain specific phenomena such as reflection, refraction, and diffraction. A simple wave is shown in Figure 3.14.*

3.20 What is the wave-particle duality? How does de Broglie's theory explain this phenomenon?

WAVE-PARTICLE DUALITY: *Is the principle, which states that both matter and energy particles exhibit some phenomena in which they behave opposite to their usual nature, that is, there are cases in which electromagnetic waves (such as lightwaves, etc.) behave as localized particles (photons) and other cases, where localized particles (such as electrons) behave as electromagnetic waves. The two aspects are being related by the de Broglie relation. de Broglie's relation is contained in the de*

Broglie's theory, which states that particles of matter have wavelike properties, which can give rise to interference effects. The wavelength (λ) associated with a moving particle, which has a mass (m) and a velocity (v) is given by:

$\lambda = h/mv$ *(3.1a)*

where "h" is the Planck's constant ($h=6.62 \times 10^{-34}$ Js) and λ is the associated wave's wavelength in free space propagation.

3.21 What is diffraction? Give an example

DIFFRACTION *is the redistribution of intensity of waves in space, which results from the presence of an object (such as a grating, consisting of narrow slits or grooves) in the path of beam of light waves. This shall split up the beam into several beams in the short range, causing interference and thus producing patterns of dark and light bands downstream (i.e., regions with variations of wave amplitude and phase).*

An example would be light hitting an edge of a wall, where its shadow has a gray region due to diffraction of light, which is light bending around the corner.

3.22 What is Heisenberg's Uncertainty principle and how does it apply to classical as well as quantum physics?

THE UNCERTAINTY PRINCIPLE: *This principle states that in any measurement of a moving particle, the product of minimum uncertainties in the value of momentum (P=mv) and position (x) is given by Planck's constant ($h=6.62 \times 10^{-34}$ Js) divided by 2π, i.e.,*

$(\Delta x \Delta P)_{min} = h/2\pi$ (3.5a)

Or,

$$(\Delta x)(\Delta P) \geq h/2\pi \qquad\qquad (3.5b)$$

where the Greek letter (Δ, Delta) represents the difference or deviation in a quantity.

Furthermore, the product of the minimum uncertainties in an energy measurement (of a moving particle) and the uncertainty in the time of measurement (i.e. the time at which the measurement was made) is given by the same constant given as above (i.e. Planck's constant divided by 2π,):

$$(\Delta E\Delta t)_{min} = h/2\pi \qquad\qquad (3.6a)$$

Or,

$$(\Delta E)(\Delta t) \geq h/2\pi \qquad\qquad (3.6b)$$

This principle states that simultaneous measurement of "position and momentum" or "energy and time of observation" are inherently inaccurate and uncertain.

Because Planck's constant (h) is a rather small number, the uncertainty of measurement of a moving particle (such as a baseball) at the classical level of observation does not pose any significant measurement error. However, at the quantum level of observation, the Heisenberg uncertainty principle (expressed by the above two equations) impose serious limitations on the accuracy of measurements of a subatomic particle (such as an electron).

3.23 How do we respond to the question of whether something is a wave or a particle?

To understand the subject of particle-wave theory and answer the question of whether something is "a particle or a wave," we need to ask a prior question, "what is the size of the created space in which the particle is placed?" The answer will solve our dilemma. Thus it is a question of size of the created space under observation that will provide the resolution to this riddle.

3.24 What does the dual theory of light state and what led to its development?

It states that:

1) Light demonstrates a wave-like property in which it behaves like an electromagnetic wave as in classical physics. Furthermore, light also has a particle-like property, in which it behaves like a localized particle of matter as in quantum physics

2) A light beam of frequency (f) is composed of a stream of photons, which are particles possessing energy (E) given by:

$E=hf$

3) The intensity of light decreases with increasing distance at any point in space, and is equal to a) The average power per unit area of the waves, or b) The density of the photons at that location.

What led to its development was based on the observation and experiments about electron beams, being a stream of particles, which have wavelike properties and behave like light beams. This observation was confirmed by electron diffraction experiments, in which electron beams would diffract when a crystal structure was placed in their path.

CHAPTER 4
Answers

QUIZ 4.1

4.1 What is the sequence of steps in the evolution of electricity?

We can see that before electricity was discovered, Man needed first to observe the phenomena, visualize the concept of electricity and then, mentally grasp how it functioned. Once that thought was visualized and grasped, then and only then came the dawn of electricity as a physical entity and its subsequent blossoming as a field of study along with its plethora of application mass as well as its many facets of existence, such as magnetism, waves, etc. This means that understanding of the basic principles, mechanisms of operation and functions of electricity actually preceded many years before the appearance of any electrical device or sophisticated machinery.

4.2 What is the first monumental discovery and why was it so important.

THE MONUMENTAL DISCOVERY #1: The Realization of Static Charge.

The importance of its discovery lies in the fact that it enabled Man to discover kinetic electricity—the most useful form of electricity.

4.3 What was the first milestone? Describe its importance.

The first milestone is the birth of static electricity and its study in a new unit of time, for the first time.

Its importance lies in the fact that when William Gilbert reported his careful and intelligent study of the phenomenon, then and only then did it begin to receive the attention which was to lay the foundation as the single most important revolutionizing agency in our today's highly evolved civilization.

4.4 What is an electrostatic machine?
Electrostatic machine is an improvement of the friction process and makes large scale generation of continuous electrostatic charges possible.

4.5 What was the first charge device?
The device invented was called a "Leyden Jar" and was the first means of storing static electricity.

4.6 What is the nature of lightning and what was the byproduct of this discovery?
By charging a Leyden Jar during a thunderstorm, Franklin and his son demonstrated that lightning and electricity (as produced by an electrostatic machine) are identical. The byproduct of this was lightning rods on roof tops to dissipate the thunder-cloud charge gradually and harmlessly to ground.

QUIZ 4.2
4.7 What was the concept behind measurement of charge?
Measurement of charge was the next step in the evolution of electricity, which made it possible to determine the nature of an unknown quantity using the known.

4.8 What is the relationship between two charges and magnitude of forces on each other?

Using the torsion balance, Coulomb found through precise measurements that "like charges" repelled each other by a force whose magnitude was proportional to the magnitudes of charges and inversely proportional to the square of separating distance between the two.

4.9 What is Coulomb's electric force law? Explain its relation to Newton's law of gravity.

"The electric forces of two electrified spheres (being either repulsion or attractive forces), is both proportional directly to the magnitudes of the charges (Q) and inversely to the square of the distance (d) between the two." Coulomb noted the similarity of his discoveries to that of Newton's "law of gravitation" involving also an inverse square-of-distance action at a distance.

4.10 What is Coulomb's magnetic force law?

The magnetic force between two magnetic poles (repulsive or attractive) is proportional to the magnitude of the poles (M) and inversely to the square of the distance (d) between the two.

4.11 Why did Coulomb discard the two-fluid theory for magnets in favor of microscopic magnetic fluid? Explain.

Coulomb discarded the macroscopic two-fluid theory and hypothesized a "microscopic magnetic fluid", which was contained in the molecule and could not be passed to other molecules. In his view, the ultimate magnetic particle was a polarized molecule and a magnet could be visualized to be made up of many polarized molecules lined up pole to pole with the magnetic effects showing up at each end called the "pole."

4.12 What field of study did the first monumental discovery lead to?

The first discovery has created a whole new field of study called "Electrostatics".

QUIZ 4.3

4.13 What is a lightning rod and how does it work? Draw a diagram.

A lightning rod is a metal rod attached above the highest part of the building and cabled from the roof to the earth and buried in a deep pit along with lots of salt (Sodium chloride).

4.14 What is the Van de Graaff generator and what is the theory of operation? Draw a simple diagram.

Van de Graaff is capable of generating electrical potentials of several million volts. The insulator belt, made of silk or rubber, passes near a charge source of several kilovolts, which sprays positive charges on the belt. The belt is moved by two pulleys. As the charges are carried into a metal dome, a metal brush takes up these charges and exchanges them for negative charges on the interior of the sphere.

4.15 What is the concept of electrostatic separation? Explain.

Electrostatic separation is used in industry to sort material that is composed of two types of granules. An example of this type of sorting is the electrostatic separation of phosphate and quartz, which are found together in raw ores.

4.16 What is electrostatic filtering and how does it work?

Electrostatic filters (also known as electrostatic precipitators) are used in industry to filter gas emissions and in homes to filter dust and pollen from the air.

4.17 How does Xerography work? Explain.

A thin coating (20-100 microns) of a photoconductive material such as amorphous selenium is put on a metal conducting plate. The parts of the photoconductor that are exposed to light become conductors, while dark regions of the photoconductor remain insulators to the flow of charge. A positively charged paper is placed on the photo-conductor and the carbon adheres to the paper, thus forming the image on the paper.

4.18 What is a cathode Ray Tube? Explain its operation.

Free electrons are produced at the cathode and are accelerated toward the anode by an electric field. Once past the anode, the thin beam of electrons is steered onto a fluorescent screen by a pair of horizontal plates and a pair of vertical plates. Rapidly varying the voltages applied to the pairs of plates result in lines and shapes being displayed on the screen by a trail of fluorescence.

4.19 What is the concept of charge separation?

Franklin developed the "one fluid" theory of electricity, which replaced the French scientist DuFay's "two fluids" description. He proposed that objects commonly exist in a balanced or neutral state. When it is rubbed by a second object and due to friction, either an excess or deficiency of charge will be created. He concluded that the process of electrification by friction did not "create" charge, but rather "separated" them.

This argument was the beginning of the "principle of conservation of charge," which was developed and expressed later into the electric Gauss's law (the law of electric charges), which states that "a quantity of electric charge can not be destroyed or created but only conserved."

4.20 What is electric bells device? Explain its operation

The mechanism of operation is based on the suspended metal clapper, which alternately contacts a bell, receives a charge and is then repelled by the like charges to swing to the other bell to repeat the same process all over again.

4.21 What is an electric wheel? Describe the mechanism of operation.

"Electric Wheel', invented by Benjamin Franklin, was one of the earliest means of producing rotary motion using electricity. Once the wheel is set in motion, the motion will perpetuate, because as the thimbles at the end of spokes move past the Leyden Jars, they are alternately attracted and repelled by the change of polarity.

CHAPTER 5
Answers

QUIZ 5.1

5.1 What was the essential milestone? Describe its importance.

The formulation and discovery of the second principle of electricity made the biggest milestone in the development of the wave theory proposed by James Clerk Maxwell decades later, which culminated and polished off the field of electricity and brought it to its full fruition as a mature subject. It was a monumental achievement in that, the second discovery made it possible for all of the future scientists to complete the later discoveries in this field and thus be able to bring this fascinating and yet vital field to full maturity.

5.2 What is the second monumental discovery and why was it so important.

THE MONUMENTAL DISCOVERY #2: *The Realization of Perpetual Kinetic Charge.*

Its importance lies in the fact that Volta had produced "an inexhaustible charge, a perpetual action, or a perpetual impulsion of the electric fluid". Although the battery's output was weaker than a Leyden jar's but it produced a continuous flow of electricity needing no outside charge. Thus it could be considered a Leyden jar with permanent charge!

5.3 What was Galvani's observation? Explain.

One day, one of his assistants accidentally touched a frog leg by a scalpel connected to a spark machine and the muscles of limb were suddenly convulsed. He set out an extended series of experiments to study and resolve the cause of the strange convulsions of frog leg specimen in response to the sparks from the machine. Later, he found out that the convulsions could also be caused by touching the frog's muscle to two dissimilar metallic probes. The intensity of convulsions varied according to the kind of metallic probes used (see Figure 5.1b). Galvani, being a physician/physical anatomist, concluded and attributed the results to animal electricity which resided in the muscles and nerves of the organism itself.

He theorized that charges flowed from the brain through nerves to the muscles. Release of stored charges by metallic contact caused the convulsions of the muscles, much like discharge of a Leyden jar.

5.4 What was the inventory of knowledge about static electricity at the time of Volta?

a) *There were positive and negative charges which represented excess or deficiency of an electric force or "fluid".*

b) *There was a mutual attraction/repulsion between unlike/like charges.*

c) *The electrostatic forces were proportional to the inverse of the square of distance between two charges.*

d) *Electric charge could be stored in a Leyden jar,*

e) *Electric charge could be produced using an electrostatic machine based on the principle of friction.*

5.5 Who was the main opponent of Galvani's theory of animal electricity and what was his explanation of the source of electricity?

When Galvani reported his findings on "animal electricity" in 1791, Volta showed great interest in his findings and set up quickly to repeat his experiments. Within a few years of investigation of this phenomenon, Volta had become an outspoken opponent of the Galvani's theory of animal electricity and his explanation of how electricity was produced by the animal's nervous system. While crediting Galvani with an original discovery, Volta disagreed with him on what produced the effects.

He proposed the theory of "bimetallic electricity". Based on experimental evidence, Volta hypothesized the source of electricity to be the result of contact of two dissimilar metals only, with the animal tissue and nerves acting merely as the indicator or conductor of electricity.

5.6 What is electromotive force and how is it related to a battery?

DEFINITION- ELECTROMOTIVE FORCE (EMF): Is the difference in electric potential that exists between two dissimilar electrodes immersed in the same electrolyte, which has many ionic particles. The difference in potential causes a flow of energy, in terms of an electric current, from the higher potential to the lower. Through his invention, Volta had shown that an electromotive force (emf) could be produced by the contact of dissimilar metals as in a battery.

5.7 What is an electric pile and how is it constructed?

By stacking pairs of metal disks (zinc and silver) in direct contact while separated by pasteboard or cloth disks soaked in saline or acidified solution, Volta constructed the first "electric pile," which was also known as "Voltaic pile". This electric pile could produce a perpetual current thus the concept of the electric battery was born!

QUIZ 5.2

5.8 How was the first electric light produced?

Sir Humphrey Davy invented the First electric light by using a bank of 500 voltaic cells connected to two pieces of carbon.   This arrangement produced a brilliant arc between the two pieces of carbon.

5.9 What is Electrokinetics? Explain its relationship to Volta's work.

ELECTROKINETICS: *Is that broad and general field of study dealing with electric charges in motion. It studies moving electric charges (such as electrons) in electric circuits and electrified particles (such as ions) in electric or magnetic fields; Volta's work opened the door to its exploration.*

5.10 What subjects does the field of Electrokinetics embrace? Give several examples.

Electrokinetics is the branch of physics that deals, in general, with moving charges and charged objects in motion. This vast field of study has many subsets including but not limited to Direct Current (DC) circuits, Alternating Current (AC) circuits, electric power circuits, electronics, electrodynamics (dealing with the interaction between electricity and a material medium, such as a conductor, etc.), so on and so forth.

5.11 What is a battery and how does it work?

An "electric battery" consists of two dissimilar metals and a liquid separator. In order to have kinetic electricity, we need to have a) The prime mover and b) The moved. The prime mover is electromotive force (emf) produced by the battery and the moved is the electric charge.

5.12 What is a condenser microphone? Explain how it works.

A condenser microphone is a capacitor consisting of a flexible plate separated by a fixed plate. The condenser microphone operates as a microphone because sound waves strike the flexible top of the condenser (also known as a capacitor) and cause it to vibrate. This flexible top is usually a polyester film coated with a thin layer of gold, which is conductive. A voltage is applied between the flexible outer plate and the inner one, causing them to be charged. The vibration due to the sound waves brings the flexible plate closer to, and farther from, the inner plate. This change of distance varies the capacitance of the condenser and produces a current.

5.13 How is chemistry and electricity intimately related to each other? Explain.

Davy showed that by sending a current through a molten salt, shiny globules of metals were isolated. Therefore, he was able to show that metal salts could be decomposed electrically into their elemental components. This was the beginning of a whole new field called "electrochemistry."

CHAPTER 6
Answers

QUIZ 6.1
6.1 What was the third discovery?
The third monumental discovery is the Realization of magnetism from electricity.

6.2 How did the second discovery lead to the third?
The second discovery paved the way to the discovery of generation of magnetism as a result of kinetic electricity, known as electric current.

6.3 What are the two types of magnets?
The two types of magnets are:
   1. *Natural Magnets, and*
   2. *Artificial Magnets.*

6.4 What are the methods of creating artificial magnets?
There are various ways of artificially inducing magnetism in un-magnetized iron or steel as follows:
   a) *Stroking or rubbing a piece of iron or steel with a strong magnet,*
   b) *Bringing a piece of iron or steel in contact or merely near a permanent magnet (not touching) and tapping it gently a number of times. This is a process commonly referred to as induction where there is no physical contact but magnetization still takes place, and*

c) *Sending a current through an insulated coil of wire, which is wound around a core (the substance to be magnetized). The electric current creates a magnetic field, which causes the core material to be magnetized. This type of magnet is called an electromagnet.*

6.5 What is the dual universe of electricity?
Electricity in terms of an electric current is always accompanied by a dual universe called magnetism; both universes coexisting inseparably and simultaneously.

QUIZ 6.2

6.6 What are the new uses for a magnetic needle? Describe each.
The magnetic needle found two new uses in the following areas:
a) The field of electric current measurement
For measurement of an electric current a magnetic needle was placed in the vicinity of a conductor. The amount of deflection of the needle of a compass corresponded to the strength of the current in the wire.
b) The field of communication or transmission of information. Using a transmission line connecting a battery (that could be switched on and off) to a remote terminal, a magnetic needle responding to a current would be able to indicate a signal from the sending end. This was the great ancestor to the electromagnetic Telegraph and then to the telephone.

6.7 How did the field of measurement take off with the help of the third discovery?

For measurement of an electric current a magnetic needle was placed in the vicinity of a conductor. This method of measurement afforded the first means of monitoring and comparing the strength of an electric current. This method of measurement was recognized by Ampere and he called this measuring electromagnetic device a "Galvanometer," which was squarely based upon the concept of interaction of electricity with magnetism. Furthermore since any form of energy can be converted into electricity, therefore its measurement relative to a known quantity was expedited.

6.8 How did the field of communication benefit from the third discovery?

Using a transmission line connecting a battery (that could be switched on and off) to a remote terminal, a magnetic needle responding to a current would be able to indicate a signal from the sending end. This was the great ancestor to the electromagnetic Telegraph and then to the telephone (to be discovered years later) and at the time promised a great advantage over the existing lantern and semaphore methods whose range was limited and subject to weather conditions. With this discovery, messages could be sent beyond the limits of human visibility.

6.9 What is a magnetic disk memory? Describe its theory of operation.

Magnetic disk memories are comprised of a layer of ferrite on a plastic or aluminum base, where the ferrite is divided into individual magnetic cells. In each cell, the ferrite is magnetically aligned in one of two directions. The cells are read and written by a read/write head.

6.10 What is a magneplane? Explain.

The magneplane is a high-speed transportation vehicle that flies above a conducting track. The magneplane operates on the principle of magnetic levitation and magnetic force. There are two parts to the operation of the magneplane: levitation and propulsion. The magneplane's levitation is caused by the repulsion of two opposing magnetic fields. The magnetic propulsion is provided due to the same principle that moves a synchronous motor. The vehicle and its coil take the place of the rotor of the synchronous motor and the track, with specially laid coils, taking the place of the stator.

6.11 What is a rail gun and describe how it works? What is its main drawback?

A rail gun operates on the magnetic expansion force found in any current-carrying loop. The parallel rails represent the "barrel" of the gun. The projectile is in contact with both rails and completes the loop. Large amounts of current are put through the loop and the result is acceleration of the projectile, because the movement of the projectile is equivalent to expanding the loop. After being accelerated by the magnetic force, the projectile leaves the "barrel" at a high velocity.

The drawback is its enormous current requirements (e.g., 300,000 A), requiring high temperatures super-conducting materials to be used in the circuit to make it practical.

CHAPTER 7
Answers

QUIZ 7.1

7.1 What is the fourth discovery?
THE MONUMENTAL DISCOVERY #4: The Realization of the Interaction of Kinetic Charge with Matter.

7.2 What is the actuality behind magnetism?
Electricity, because it can produce magnetism on a macroscopic level (by a current through a wire) or on a microscopic level (on a molecular or atomic level).

7.3 What is Ampere's contribution to the field of electricity?
The experimental investigation by which Ampere established the laws of circulating current and the mechanical action between electric currents as well as the emerging field of electrodynamics were his contributions. These led Maxwell to describe Ampere as the "Newton of electricity".

7.4 What is Ampere's force equation?
Ampere's theory of electric currents like the Newtonian theory of gravitation was also one of instantaneous action at a distance. The electromagnetic force between electric currents was proportional to the magnitude of the current elements and inversely to the square of their separation.

7.5 Which is more senior: electricity or magnetism? Why?

Electricity! Ampere knew that Electricity and magnetism were related and one did not exist without the other but the question he faced was which one was more fundamental? Breaking down the magnetic properties to a circulating current (either on macroscopic or microscopic scale) he concluded that electricity was more fundamental.

7.6 What is Ampere's circuital law?
Ampere's Circuital Law: *States that the magnitude of the magnetic field integrated in any closed path (or loop) around a conductor is equal to the current enclosed in the loop.*

QUIZ 7.2
7.7 What was the first alternator? Describe its parts.
The first alternator was designed and built by Hippolyte Pixii, an instrument maker in Paris, France, in 1832. A horseshoe magnet mounted on a vertical axis could be rotated over the tips of a U-shaped iron armature with two wire wound coils. When the horseshoe magnet was revolved, the alternating north and south poles passing over the coils and cutting them at 90° angle, generated an alternating current (AC).

7.8 What is an electric dynamo? Describe its operation.
The dynamo is basically an alternator in which the generated alternating current is converted by a commutator to a direct-current output. By replacing the commutator with collector-rings, the machine can be converted to an alternating current (AC) output.

7.9 What was the first Electromagnetic motor? Draw a diagram and show how it works.

The first electromagnetic motor was ingeniously designed and built by Faraday in 1821. On the right side of Figure 7.7, a magnet was fixed in an upright position in a cup of mercury, with one pole of the magnet exposed. A suspended wire, with its lower end in the pool of mercury, free to circle around the magnet, was thus connected to one terminal of a battery via mercury cup. The other end of the wire was dipped into another mercury cup (left side) allowing connection to the other terminal of the battery thus creating a closed circuit loop. The magnet rod, on the left, was pivoted in a mercury cup and was allowed to rotate around the wire.

When the battery action was engaged in the circuit, the free end of the wire (right side) whirled around the magnet as long as the current was maintained, while magnet rod on the left rotated around the conducting wire in the mercury cup. Both rotations were due to the interaction between the magnetic field around the wire and the field of the magnet.

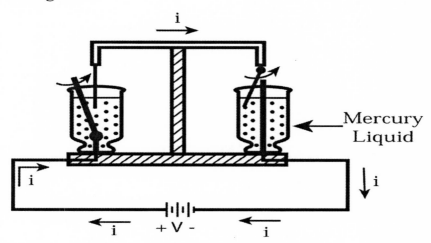

FIGURE 7.7 The first electromagnetic motor.

7.10 In general, what type of applications can the principles of electrodynamics be used for?

Ampere's development of the subject of electrodynamics opened a whole new vista of new applications of electric currents (such us telegraph, electric motors, electromagnetic generators, etc.) and brought about a resurgence of ideas in the scientific community. The new theories of the electric circuits broadened and revolutionized the scope of electricity, which was dominated by the electrostatics for so many years.

CHAPTER 8
Answers

QUIZ 8.1
8.1 What is the definition of field?

DEFINITION-FIELD: *An invisible entity that is distributed over a region of space and whose properties are a function of space and time. It acts as an intermediary agent in interactions between energy and matter particles.*

8.2 Why is the Newtonian theory deficient in the field of electricity? What are the shortcomings of this theory with regard to electricity?

The Newtonian explanation of electromagnetism had several flaws and deficiencies and thus was not able to explain the observed phenomena fully.

    a) Electromagnetism was in contradiction with the Newtonian concept of forces acting in a straight line (or direct-action forces), since the electromagnetic forces had a direction with a right angle relationship between the current and the resulting magnetism.

    b) Newtonian forces were instantaneous and thus in contradiction of the electromagnetic forces which implied a time factor directly associated with the distance between the source location and the receipt point.

8.3 What was the Newtonian misconception?

The Newtonian explanation of electromagnetism had several flaws and deficiencies and thus was not able to explain the observed phenomena fully. The points of discrepancy could be summed up as:

a) The electromagnetic forces had a direction with a right angle relationship between the current and the resulting magnetism.   This fact introduced the possibility of rotation.   This was in contradiction with the Newtonian concept of forces acting in a straight line (or direct-action forces).

b) The electromagnetic forces implied the possibility of a time factor, which was directly associated with the distance between the source location and the receipt point.

8.4 What is the fifth monumental discovery?

THE MONUMENTAL DISCOVERY #5: *The Realization of Force Fields.*

8.5 What is meant by the universal nature of force?

Faraday had a philosophical approach to his work and believed in the "unity of physical phenomena", in particular, he considered that matter, electricity and light were of the same origin. Faraday's philosophical considerations about matter, electricity, and light conform to the modern discoveries, since we know today that these are all derived from "energy", each having progressively a higher frequency with a corresponding decreasing wavelength.

8.6 What are lines of force?

DEFINITION-LINE OF FORCE (ALSO KNOWN AS FLUX LINE OR LINE OF FLUX): *Is an imaginary line in a field of force (such as electric field, etc.), whose tangent at any point provides the direction of the field at that point; The lines are spaced so that the number of lines through a unit area perpendicular to the field represents the intensity of the field.*

QUIZ 8.2

8.7 What is the principle of conservation of charge and what led to its development?

THE PRINCIPLE OF CONSERVATION OF CHARGE: *This principle states that in any closed system existing in the physical universe, charge can neither be created nor destroyed but only separated and thus is conserved.*

8.8 What was Faraday's prediction well in advance of its discovery?

Using lines of force theory, Faraday was able to theorize well in advance of its time, the future of electromagnetics as being able to radiate out into space. He considered radiation to be a category of vibrations along the lines of force.

8.9 Give a few examples of force fields.

Electricity, magnetism, gravity, sound, etc.

8.10 What is a transformer? Explain its theory of operation.

A transformer consists of two windings: Primary winding A, and secondary winding B with a magnetic core connecting the two. When winding (A) is energized by the battery, the galvanometer needle connected to winding (B), would spin around several times. The needle would soon come to rest and become indifferent to the steady current. The needle responds similarly but in a reversed fashion when the battery is disconnected. The rise and fall of magnetism in the iron core of the ring, with the making and breaking of the current to winding (A), induces a current in winding (B) as indicated by the galvanometer (G). "Change of magnetism" is essential for induction and transformer action.

8.11 What is an electromagnetic motor? Draw a diagram and show how it works.

It was understood that when wires move so as to cut a magnetic field, a power is called into action, which tends to urge an electric current through them. Moreover, experiment suggested that an iron core was better than that of air, wood, etc., because it concentrated the magnetism for the most effective induction of electricity. It was also found that when the magnetic field directly cuts across at 90° angle with the conductor, maximum current was induced. These observations led to the development of the electromagnetic motors which includes the alternator, the dynamos, etc.

A modern and yet very typical example is drawn in the Figure 7.6.

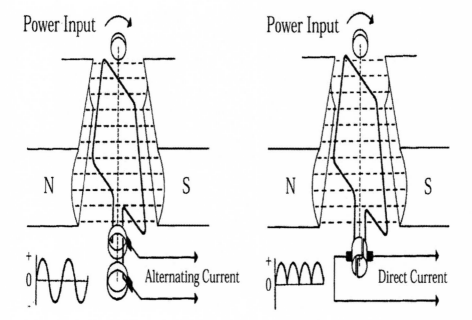

FIGURE 7.6 An electric dynamo.

8.12 What is Arago's disk? How does it work and what is the main operating principle here?

Whenever an enclosed copper disk was rotated by the diving pulley inside an enclosure, the compass needle outside the enclosure would spin around. Many scientists had labored at various explanations for this strange action, but it was Faraday who finally resolved the mystery. Faraday's discovery of electromagnetic induction revealed that rotation of the compass needle was the result of interaction between the magnetic needle and the disk. The magnetic needle induced a current in the moving disk, which in turn set up a magnetic field. This induced magnetic field (caused by the induced current in the disk) would now interact with the magnetic needle and thus causing it to spin.

QUIZ 8.3

8.13 How does Earth's ionosphere interact with radio frequency signals?

The earth's ionosphere, composed of plasma, starts at roughly 80 km above the earth's surface. This plasma is formed from air molecules that are excited by ultraviolet radiation from the sun. It has been found that for frequencies lower than 9 MHz, the ionosphere will reflect electromagnetic radiation. At frequencies above 9 MHz, the ionosphere allows electromagnetic radiation to pass through with very little reflection.

8.14 What is antenna orientation and how does it affect signal reception?

*The orientation of the **electric field (called polarization)** in a broadcasted signal determines the best antenna configuration to receive the signal.*

8.15 How can we double the communication capacity by using polarization?

Because of the popularity of satellite use, the frequency bands for satellites are crowded, In order to increase the amount of information that can be transmitted in a given band, satellite systems transmit two distinct signals polarized 90° apart. By using such orthogonal polarization, the satellite can transmit twice the amount of information in a given bandwidth with no interference between signals.

8.16 What is a grating reflector and how does it work?

Reflection of a signal with a grating is possible because when a polarized wave strikes a surface, there is only surface current flowing in the direction of the polarization of the electric field. Thus, if the incident wave is polarized with its E-field in the y-axis, a grating of wires in the y-direction would effectively reflect this signal.

QUIZ 8.4

8.17 What is a frequency meter and what can it be used for?

In a frequency meter a cavity is coupled to a waveguide through a small opening. A micrometer adjusts the length of the cavity by moving its free end to cause resonance. A probe at the end of the waveguide senses the power of the unknown frequency signal.

A frequency meter uses the fact that a conducting metal cavity of the right length will cause an electromagnetic wave inside the cavity to become a standing wave, which really means that a resonance will occur.

8.18 How is the wavelength of a signal measured? Explain.

a) Method #1: To find the signal frequency, the micrometer is adjusted until the detector registers a decrease in power. This decrease indicates that the signal is resonant in the cavity and it is drawing power to maintain this resonance. The micrometer reading is then compared to a chart to determine the signal frequency. Once the frequency (f) is known we use: $\lambda = c/f$, to find the wavelength (λ).

b) Method #2: The wavelength of the electromagnetic wave is measured by sending the signal in a special coaxial cable. This cable has one end shorted together by a conducting annular ring. The result is that a standing wave is produced along the coaxial cable. If a lengthwise slit is made in the cable, a probe can be inserted that will measure the power of standing wave. As the probe is moved along the cable, a standing wave pattern will be registered. The wavelength of the unknown electromagnetic wave will be twice the wavelength of the standing wave. The standing wave is measured using the probe and will have a reading as a result of the combination of two waves: a) Incident wave and b) Reflected wave.

8.19 Why the concept of fields has revolutionized many areas of study?

The subject of fields, even though primarily proposed by Faraday for electricity and magnetism, has actually spread out to other subjects (such as Gravity, Fluid Dynamics, etc.), and thus has made the respective scientists, able to inject and integrate the concept of fields into their respective subjects, all to make them more practical and closer to truth. It has even spread to such distant fields as spiritualism and mental phenomena, where research has shown that there is an electric field surrounding the human body which can be influenced by the spirit or mind.

8.20 What is meant by basal unity in nature?

*Faraday's vision was based on the oldest and yet continuing quest in the philosophy of science-- the **Basal Unity in Nature** i.e. all forces or energy sources come from a common source and therefore are convertible into one another.*

CHAPTER 9
Answers

QUIZ 9.1

9.1 What is the definition of a wave?

DEFINITION-WAVE: *By wave we mean a disturbance that propagates from one point in a medium to other points without giving the medium as a whole, any permanent displacement.*

9.2 What is the sixth monumental discovery?

THE MONUMENTAL DISCOVERY #6: The Realization of Electronic Waves.

9.3 Why is "waves" a subset of kinetic energy?

Waves are flows and are actually energy in motion. Therefore, we see that a subset of kinetic energy is the field of "waves" of which "electromagnetic waves" is but a small subset.

9.4 What did Maxwell's physical model unravel?

Using his model, he made several discoveries including:

a) *The electromotive force on any element of a conductor is measured by the instantaneous rate of change of magnetic flux on that element. This change of magnetic flux could be either in magnitude, direction or both.*

b) *Placing a dielectric in an electrostatic field causes strain in the direction of electric force. This strain effect, which he called polarization, was actually the internal displacement of bound charges in the dielectric medium due to the electric pressure.   This improved the overall charge storage capability of the device.*

c) *The displacement current is capable of creating a magnetic field, exactly the same way as conduction current is.*

d) *Electromagnetic waves require space and time for propagation, which imply a velocity.   The velocity was calculated to be approximately 3.0 x 10^8 m/s, which was in exact agreement with the velocity of light already known by optical experiments.   This was the prelude to understanding radio waves purely based on observational evidence without any proof. This theoretical prediction was actually man's first realization that radio waves even existed, let alone using them for communication and so many other life enhancing benefits.*

9.5 What is meant by the "displacement current"? What was its importance?

Maxwell used his physical model to extend his investigations beyond the conduction current (I_C) to derive the behavior of electricity in a dielectric (such as air). He hypothesized that applying an electric pressure (i.e., a voltage) on a dielectric produced a displacement current (I_D) with an accompanying magnetic field that traversed the dielectric space. The dielectric space had become an active part of the electric circuit. In this fashion the continuity of current was preserved and the total current consisted of two parts: a) the conduction current and b) the displacement current.

With the addition of this term, the insulating medium of the condenser (such as air, ceramic, etc.) could be considered a conductor and contributed to the current flow in the circuit. With this change in the equations, Maxwell had miraculously achieved a unified set of four self-consisting equations, which could describe any electric, magnetic or electromagnetic phenomena. In short, he had hit pay dirt!

9.6 What was Maxwell's contribution to the field of electromagnetics?

Maxwell's physical model had now delivered two extraordinary discoveries that were highlights of his paper entitled "on physical lines of force". The first was a new kind of current called "displacement current" and the other was a hypothetical undulating source that related light and electromagnetic phenomena by a common velocity.   With these two achievements, Maxwell proceeded to build the unified theory of electricity and magnetism.

QUIZ 9.2

9.7 How many and what were the original Maxwell's equations?

Maxwell's eight general equations were listed in his paper "dynamical theory of the electromagnetic field" and, in Maxwell's own words, expressed the following relationships:

a) EQUATION #1: The relation between electric displacement current, true conduction current and total current compounded of both,

b) EQUATION #2: The relation between the lines of magnetic force and the inductive coefficients of a circuit as already deduced from the laws of induction,

c) EQUATION #3: The relation between the strength of a current and its magnetic effects according to the electromagnetic system of measurement,

d) EQUATION #4: The value of the electromotive force in a body as arising from the motion of the body in a field, the alteration of the field itself and the variation of electric potential from one part of the field to another,

e) EQUATION #5: The relation between an electric current and the electromotive force which produces it,

f) EQUATION #6: The relation between electric displacement and the electromotive force which produces it,

g) EQUATION #7: The relation between the amount of free electricity at any point and the electric displacements in the neighborhood,

h) EQUATION #8: The relation between the increase or diminution of free electricity and the electric currents in the neighborhood.

9.8 How many and what are the modified Maxwell's equations?

Later the eight equations were revised and re-ordered by Maxwell and was presented in a paper entitled "treaties on electricity and magnetism". The equations became four in number and are traditionally called "The Maxwell's equations", which is one of the pivotal and generalized theories of physics.

At first, Maxwell used partial differential equations for his formulas, but later used vector calculus and employed "Divergence (Div)", "Curl" and "Gradient" of vector fields to address these four equations in a compact format. The four modified and new equations are:

EQUATION I: *Current flowing in a wire generates a magnetic flux that encircles the wire in a clockwise direction when the current is moving away from the observer. This is known as "Ampere's Law".*

EQUATION II: *When a magnetic field cuts a conductor, or when a conductor cuts a magnetic field, an electrical current will flow through the conductor if a closed path is provided over which the current can circulate. This is known as "Faraday's Law".*

EQUATION III: *The summation of the normal component of the electrical displacement vector over any closed surface is equal to the electric charge within the surface, which in essence means that the source of the electric flux lines is the electric charge. This is known as "Electric Gauss's Law".*

EQUATION IV: *The summation of the normal component of the magnetic flux density vector over any closed surface is equal to zero, which in essence means that the magnetic flux lines have no source or origin. This is known as "Magnetic Gauss's Law".*

9.9 What was one of the greatest unifications in physics? Explain.

Maxwell was on the track of one of the great unification in physics: light and EM waves.

9.10 What were the four factors crucial in the unification process of light and EM waves?

The correlation between the two depended on several factors including:

a) *The two being the same types of waves,*
b) *The two having equal velocity,*
c) *The two having the same propagating properties when transmitted through the same medium,*
d) *The two being polarized in their vibrations.*

9.11 What made the concept of Newtonian mechanics obsolete?

The inclusion of light into the realm of electromagnetics, made the Newtonian mechanics (which was "instantaneous action at a distance" type theory) less tangible and obsolete. The time factor was implicit in wave propagation. This meant that from Maxwell's force field point of view an electromagnetic disturbance, being received at a point in space a distance away from the source point, has a time-delay response built into it.

9.12 What is ether and why was it necessary to include it in the scientific thinking in the 19th century?

The notion of the universal medium goes back to the Greeks who had a word for it --the ether. From their point of view, it filled all space of the starry heaven and all planets moved in it. Upon investigation of optics in 17th century, the idea of ether was brought to the forefront as a way of explaining the transport of light, very similar to the way a gas such as air, carries sound waves. With the introduction of Maxwell's equations in 1884 showing that light was also an electromagnetic phenomenon, the ether became more essential. Thus Maxwell hypothesized that the electromagnetic fields or waves propagated in a medium, which existed everywhere. Ether was described by Maxwell as a necessary transport medium for EM waves including light.

QUIZ 9.3

9.13 What experiments led to the discovery of producing electricity from magnetism?

EXPERIMENT #1: *A magnetic bar was plunged into a coil of wire, which produced an electromotive force causing a current in the coil of wire. This case corresponded to a moving magnetic field and a static conductor.*

EXPERIMENT #2: *A rotating copper disk was inserted in the field of a stationary magnet, which produced an electromotive force causing varying circular currents in the disk. This case corresponded to a static magnetic field and a moving conductor.*

9.14 What are the two steps leading to the concept of "producing electricity from magnetism"?

The two steps are:
   a) The induced electromotive force causes a current that in turn generates (or induces) a magnetic field.
   b) The induced magnetic field will oppose the original magnetic field, thus causing a braking action on the moving part.

9.15 Describe what is meant by remote sensing?

Remote sensing allows aircraft or spacecraft to detect and distinguish features and terrain on the earth's surface.

Microwaves are used for remote sensing. There are two types of remote sensing configurations: active and passive.

9.16 How does nearby thunderstorms interfere with radio signals?

Nearby thunderstorms cause an increase in the noise and static in an AM radio. The reason for increased noise is the electric field, which is caused by the frequent lightning strikes that make up a thunderstorm.

QUIZ 9.4

9.17 How does a microwave reflector operate?

A large conducting plate can be effectively used to reflect microwave beams. The criterion for effective reflection is that the plate must be large compared a) to the wavelength of the signal and b) width of the signal beam.

Microwave reflectors are useful because they allow antennas to be mounted on the ground instead of on high receiving or relay towers. The reflectors are mounted high on the towers and reflect the microwave signal into the antenna on the ground.

9.18 What is the concept of polarized light?

By polarization we mean the direction of the E-field as a function of time. The sky polarization depends on the angle between the sun's rays to a point (P) in the sky and the observer's line of sight to the point (P). Because of scattering from air molecules, sunlight is completely unpolarized when an observer looks directly toward the sun and becomes more linearly polarized when the observer looks at other parts of sky farther away from the direct path to the sun.

9.19 What is meant by glareless headlight for cars?

In nighttime driving, vision can be greatly impaired by oncoming headlights. The glareless headlight system is composed of headlights that are polarized at an angle of 45° and a windshield polarized at the same angle. In this system un-polarized headlight glare is screened out by the polarized windshield, while the headlight rays are still allowed to pass through. In this way, driver visibility is maintained and glare is reduced.

9.20 What is a stereoscopic picture?

Stereoscopic pictures are well-known for their ability to give a three-dimensional effect from a two-dimensional print. To produce a three-dimensional (3-D) image, the human mind uses a slightly different view from each eye. Stereoscopic pictures produce the 3-D effect by giving each eye an image that is polarized differently. The pictures are made using two polarizations and then are viewed with glasses that have a differently polarized lens for each eye.

9.21 What is a Brewster window?

Brewster angle is defined to be the angle of incidence of light reflected from a dielectric surface at which the reflection coefficient becomes zero when the light's electrical field vector lies in the plane of incidence (i.e., parallel polarization case). In other words, if a parallel polarized wave is incident at a dielectric surface at the Brewster angle, all of the wave will be transmitted through and there will be no reflection. Generally speaking, the concept of Brewster angle applies to any electronic wave of any frequency, not just light waves. Many gas lasers use a Brewster window to produce polarized light at their outputs.

9.22 Draw a diagram showing how charge leads to the universe of waves?

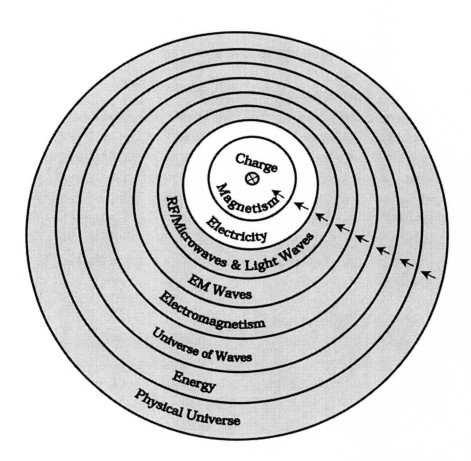

FIGURE 9.16 A diagram showing the underlying subsets of the universe of waves and how it connects to the physical universe.

CHAPTER 10
Answers

QUIZ 10.1
10.1 What is Maxwell's legacy?

The sixth principle of electricity was uncovered on a theoretical basis but its workability was uncertain since no proof, indicating its truth, existed at that time. The sixth principle completed the series of breath-taking discoveries, which had begun with William Gilbert and now actually had ended with Maxwell. The complete set of discoveries led to the four Maxwell's equations, which provided the cornerstone of all modern electrical innovations and inventions.

However, in this set of monumental discoveries, the sixth principle was the most dynamic of all, since if proven, would lead to a manmade universe of electronic waves and open up a whole gamut of applications never envisioned before.

10.2 What are the two questions that Maxwell failed to address?

Maxwell had arrived at the concept that oscillating electrical disturbances proceeding in free space should be in the form of electrical and magnetic waves traveling with a finite velocity equal to the speed of light.

Unfortunately, Maxwell did not offer any method by which the propagation of light waves or (EM waves) of much longer wavelength, which he assumed to exist at the time could be proven experimentally. His short life left to his successors a) the burden of demonstration and proof of existence of EM waves and b) their relationship to light.

10.3 What were the three experimental contributions that Hertz made to the field of electromagnetism?
a) Generation of EM waves,
b) Transmission of waves and wave speed in air, and
c) Wave reflection and formation of standing waves.

10.4 What were the four properties that were examined for verification of EM waves vs. light waves? Describe each.
There are a total of four areas that need to be investigated when dealing with the wave phenomena:

a) Reflection,
b) Rectilinear propagation,
c) Polarization and
d) Refraction.

These four areas altogether were distinct characteristics of light and radiant heat at that time and the exact experiments to verify the above four characteristics were an important step in the direction of identifying light with EM waves.

10.5 What has made Hertz and Maxwell two inseparable names in the field of electromagnetism?
In their two short lives: a) Maxwell achieved a purely theoretical prediction of the sixth principle that electromagnetic energy could propagate as waves and b) Hertz had shown experimentally the truth of that prediction.

QUIZ 10.2

10.6 What was Edison's DC system composed of?

This system was proposed by Thomas Edison and was use of DC power to energize the household items.

10.7 What was the disadvantage of the Edison's DC system?

The one major disadvantage was that every increase of the load at the customer caused an increase in current. Serving high-density load areas required very large currents using huge copper conductors. Then there was the problem of undesirable heating losses in the conductors, which increased with the square of current. Furthermore, there was a drop in voltage due to resistance of conductors that could dim lamp illumination. As a result, the distances of distribution was limited and localized to about one square mile per powerhouse.

10.8 What principle was Tesla's power distribution system based upon and what did it consist of?

Tesla's AC system was based on the alternating current (AC) and had the unique advantage of raising or lowering the voltage of a power line through the use of a simple stationary device called a "transformer". The transformer consisted of primary and secondary coils, which was separated electrically but joined magnetically.

10.9 Why did Tesla's AC system won over the Edison's DC system? Draw a diagram to show.

The use of transformers gave AC power system a formidable advantage over DC power systems, which could not use the transformer principle. Electricity at the generation point could be stepped up to thousands of Volts

(Nowadays, millions of Volts) with the corresponding reduction of current and hence smaller wire size, ready for long-distance transmission. Then, at the point of use, the voltage could be transformed down to the required value. One power plant would serve a greatly expanded area of utilization. This power transmission system proved successful by overcoming the distance limitation of a DC system.

A typical AC power transmission line is shown below.

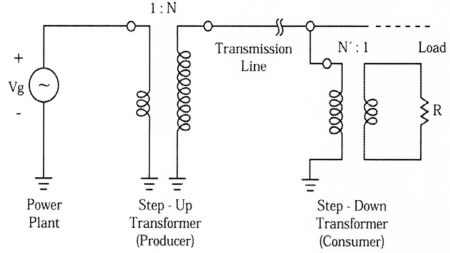

CHAPTER 11
Answers

QUIZ 11.1

11.1 What are the supplemental discoveries?

There are several subordinate discoveries, which help explain certain phenomena particularly in circuit analysis and design. These subordinate discoveries are ordinarily referred to as "supplemental discoveries".

These discoveries were realized all in the nineteenth century and have had tremendous influence in shaping the way electrical engineering is taught or practiced today.

These supplemental discoveries put the finishing touches on the six monumental discoveries and fill in the gaps that may have been left behind in this series of high speed discoveries of enormous proportions. We could liken the four postulates of physics (two workable and two implicit) as the concrete foundation of a building, the six monumental discoveries as the actual framework and skeleton of the building with the supplemental discoveries as the fixtures and furnishing to make it a functional building, and the science of electrical engineering as one of the most exciting fields of study.

11.2 What is meant by the first supplemental discovery?

THE FIRST SUPPLEMENTAL DISCOVERY: *Electric charge is the source of electric field.*

11.3 What is the Gauss's law of electrostatics?

GAUSS'S LAW OF ELECTROSTATICS (ALSO CALLED ELECTRIC GAUSS'S LAW): This law states that the total flux of electric displacement vector (D) emanating from any closed hypothetical surface [called a Gaussian surface (S)] equals the net amount of electrostatic localized charge (Q) enclosed by the surface.

11.4 What is the dual of the first supplemental discovery?

THE MAGNETIC DUAL OF THE FIRST SUPPLEMENTAL DISCOVERY: Circulating magnetic flux lines close on themselves thus have no divergence and result in no magnetic charge.

11.5 What is the magnetic Gauss's law?

MAGNETIC GAUSS'S LAW: Asserts that the net magnetic flux (Φ) through any closed Gaussian surface (S) is always zero.

QUIZ 11.2

11.6 What is the second supplemental discovery?

THE SECOND SUPPLEMENTAL DISCOVERY: There is a linear relationship between a voltage (V) applied across a resistive material and the resulting current (I) through it. The ratio of the voltage to current is termed "resistance".

11.7 Define Ohm's law and give an example of its application.

OHM'S LAW: The current value is the algebraic sum of all electromotive forces in the circuit divided by the total resistance, i.e.,

$$I = \Sigma(emf)/\Sigma(R) \qquad\qquad (11.1)$$

11.8 What does the resistance of a piece of wire depend on?

Resistance (R) is given by:

$$R = \frac{\ell}{\sigma A} \qquad (11.3)$$

where σ, ℓ and A are the conductivity, length and the cross sectional area of the material.

11.9 What similar field of observation Ohm used to pattern after and develop his famous law?

Ohm's work was established upon an observation he had made earlier about the theory of heat flow in a material as presented in 1822 by Joseph Fourier (1768-1830), the French mathematician and physicist. In his work Fourier had discussed how heat transferred from a higher temperature to a lower one. He stated that the quantity of heat flowing in a given direction would be the product of: a) The conducting area normal to the path, b) The temperature gradient along the path, and c) Thermal conductivity.

11.10 How does Ohm's law apply to the microscopic world?

Ohm's law as represented by Equation 11.3, applies to the macroscopic world where the lengths involved are noticeably large. However, on a microscopic scale the Ohm's law at each point in a material under the influence of an electric field (E), can be written as:

$$J = \sigma E \qquad (11.4)$$

Where J is the current density given by:

$$J = I/A \qquad (11.5)$$

and A is the cross sectional area of the material. Thus we can see that Equation 11.4 represents the differential form of Ohm's law.

QUIZ 11.3

11.11 How does temperature affect the resistance of a
        material?

Ohm could not have ever visualized a conductor without resistance. However, he discovered that the resistance of a material was influenced by the temperature. This recognition was the first step on a road that culminated in the achievement of zero resistance, nearly a century later. Thus the concept of superconductivity was born.

11.12 What is a superconductor? What are some possible
        applications for it?

Other metals entered the state of zero resistance at a temperature called "critical temperature". With this discovery, Onnes realized that superconductors, having a zero resistance, could carry large amounts of current, and thus were accompanied by large magnetic fields without the usual heating effects that accompanied non-superconductors.

A plethora of applications of superconductive metals and alloys has been envisioned and is currently under research and development. These are in the area of electric power generation and transmission, in high energy particle accelerators, in Magneto-hydrodynamics (the science dealing with interaction of a magnetic field and a conducting fluid such as ionized gas, liquid metal, etc.), in Microwaves, in Nuclear fusion systems, in vehicular transportation (such as magnetic levitation train or magneplane, etc.) and others.

11.13 What is the dual of the second supplemental discovery?

THE MAGNETIC DUAL OF THE SECOND SUPPLEMENTAL DISCOVERY: *There is a linear relationship between the flux lines of the magnetic field and the Magnetomotive force (mmf).*

11.14 Define the magnetic Ohm's law.

MAGNETIC OHM'S LAW: *The magnetic flux (Φ) value is the algebraic sum of all magnetomotive forces (mmfs) in the circuit divided by the total reluctance ($\mathfrak{R}$), i.e.,*

$$\Phi = \Sigma(mmf) \, / \, \Sigma(\mathfrak{R}) \tag{11.9}$$

Or, if we designate " $\Sigma(mmf)$" by "F_{tot}", and $\Sigma(\mathfrak{R})$ by $\mathfrak{R}_{tot}$, we can write:

$$F_{tot} = \mathfrak{R}_{tot} \, \Phi \tag{11.10}$$

11.15 What is meant by magnetomotive force (mmf)?

Magnetomotive Force "mmf" being the dual of Electromotive Force (emf), is the cause of magnetic flux(Φ) circulation and is given by:

$$mmf = NI \tag{11.11}$$

Where N is the number of turns in a coil and I is the current in the coil.

QUIZ 11.4

11.16 What is the third supplemental discovery?

THE THIRD SUPPLEMENTAL DISCOVERY: *Realization of Complex number algebra as a mathematical tool in analyzing and designing circuits dealing with sinusoidal signals or alternating currents.*

11.17 What are the main laws pertaining to all AC circuits? Briefly describe it.

THE GENERALIZED OHM'S LAW: *Under steady state conditions, for a circuit consisting of a resistor (R) in series with a capacitor (C) and an inductor (L), the applied voltage can be written as:*

$$V=ZI, \tag{11.13}$$

Where,

$$Z=R+j(\omega L-1/\omega C), \quad \omega=2\pi f \tag{11.14}$$

11.18 What is the concept of impedance and how is it used to solve AC circuits?

The concept of impedance is similar to resistance, except that it is no longer a constant real value, but a complex number whose value depends on the frequency of operation (f). The inverse of impedance is called admittance (Y) given by:

In general, the impedance (Z), being a complex number, has a real part (called resistance, R) and an imaginary part (called reactance, X), thus we have: $Z=R+jX$

To solve any AC circuits we use the Generalized Ohm's law.

11.19 What is the fourth supplemental discovery?

THE FOURTH SUPPLEMENTAL DISCOVERY: *All current-carrying electrical wires exhibit self induction when the current is changed in magnitude or direction. The self induction impedes change in the current flow (magnitude and direction) and thus causes a voltage drop.*

11.20 If a running electric current suddenly is interrupted, what causes the sparks to fly off and what is this phenomenon called?

Henry had noticed that when a short wire is connected across a battery in series with a switch, no spark is perceived when the switch was closed or opened. However, if a relatively long length of wire (30-40 ft.) was used instead, the situation was quite different. There were no sparks perceptible when the switch was closed, but when the switched was opened a noticeable spark was present. The spark effect was more pronounced if the long length of wire was turned into a coil or helix.

He interpreted the spark at the point of interruption as due to the sudden collapse of the magnetic field around the conductor itself, which generated a reverse voltage sufficient to make the spark. This phenomenon was called "self-induction". Self-induction is a property of all electric circuits, which is always present to a greater or lesser degree, and can be measured in terms of a parameter called "inductance".

11.21 What is the definition of coefficient of self induction (or inductance) as symbolized by "L"?

An important relationship can be derived by noting that the magnetic flux (Φ) is proportional to the current:

$$N\Phi = LI \qquad (11.15)$$

where L is the coefficient of self-induction.

Differentiating both sides we obtain the induced voltage (V) as:

$V=d(N\Phi)/dt$ (11.16)

Or,

$V=LdI/dt$ (11.17)

The induced voltage (V) opposes the original applied emf and thus appears as a voltage drop in the circuit.

QUIZ 11.5

11.22 What is the fifth supplemental discovery?

THE FIFTH SUPPLEMENTAL DISCOVERY: *In the field of electricity, there is no "absolute electric charge", that is to say, no single electric charge can exist all by itself (see Chapter 1, section 1.6.4, The Principle of Relativity of knowledge).*

11.23 What is the corollary 5a to the fifth supplemental discovery?

COROLLARY 5A-THE PRINCIPLE OF CONSERVATION OF CHARGE: *Electric charge can neither be generated nor destroyed. That is to say, it can only be separated into its opposite charged components.*

11.24 What is the corollary 5b to the fifth supplemental discovery?

COROLLARY 5B- THE PRINCIPLE OF CHARGE MIGRATION: *Static charges will migrate to the outer surface of a perfect conductor and will reside there regardless of where they are placed.*

11.25 What is meant by the sixth supplemental discovery?

SIXTH SUPPLEMENTAL DISCOVERY: *When an external electric field is applied to a dielectric material, a molecular polarization takes place microscopically, which alters the electric field's density of flux lines inside and thus causes charge storage effects.*

11.26 What is the dual of the sixth supplemental discovery?

THE MAGNETIC DUAL OF THE SIXTH SUPPLEMENTAL DISCOVERY: *When an external magnetic field is applied to a magnetic material, the material becomes magnetically polarized, which alters the density of the magnetic flux lines inside. If the alteration is an enhancement of the internal magnetic field, the material is called paramagnetic; whereas if the field is de-intensified, the material is diamagnetic.*

QUIZ 11.6

11.27 What is the seventh supplemental discovery?

THE SEVENTH SUPPLEMENTAL DISCOVERY (ALSO CALLED THE PRINCIPLE OF CONSERVATION OF CURRENT): *Electric current in a circuit can neither be created nor be destroyed. Thus the algebraic sum of currents crossing any closed surface, containing any part of an electrical circuit, remains zero.*

11.28 What law is a direct byproduct of seventh supplemental discovery? Describe it.

ELECTRIC KIRCHHOFF'S CURRENT LAW (EKCL): *In any lumped-element network and at each instant of time, the algebraic sum of the currents at each node must be equal to zero, i.e.,*

$$\sum_{n=1}^{N} I_n(t) = 0 \qquad\qquad (11.18)$$

Where N is the total number of branches connected to any given node and $I_n(t)$ is the current in the nth branch.

11.29 What is the dual of the seventh supplemental discovery?

THE MAGNETIC DUAL OF THE SEVENTH SUPPLEMENTAL DISCOVERY: *Magnetic flux can neither be generated nor be destroyed. Thus the algebraic sum of fluxes crossing any closed surface, containing any part of a magnetic circuit, remains zero.*

11.30 Describe the law that is directly derived from the dual of the seventh supplemental discovery?

MAGNETIC KIRCHHOFF'S CURRENT LAW (MKCL): *In any magnetic network and at each instant of time, the algebraic sum of the magnetic flux at each node must be equal to zero.*

$$\sum_{i=1}^{N} \Phi_i(t) = 0 \qquad\qquad (11.19)$$

Where N is the total number of branches connected to any given node and $\Phi_i(t)$ is the magnetic flux in the ith branch.

QUIZ 11.7

11.31 What is the eighth supplemental discovery?

THE EIGHTH SUPPLEMENTAL DISCOVERY (ALSO CALLED THE PRINCIPLE OF CONSERVATION OF VOLTAGE): *Electric voltage can be neither generated nor destroyed in a conservative electric field (such as static field or quasi-*

static field). In other words, traveling on a closed path, we gain a zero net voltage.

11.32 What is the electric Kirchhoff's voltage law?

ELECTRIC KIRCHHOFF'S CURRENT LAW (EKVL): *In any lumped-element network and at each instant of time, the algebraic sum of the branch voltages around a closed loop must be equal to zero.*

$$\sum_{m=1}^{M} V_m(t) = 0 \qquad (11.20)$$

Where M is the total number of branches in the loop and $V_m(t)$ is the branch voltage in the mth branch.

We can write this equation more explicitly as

$$\sum_{k=1}^{M_1} emf_k = \sum_{i=1}^{M_2} V_i \qquad (11.21)$$

Where emf_k is the voltage of the kth sources (e.g. a battery) and V_i is the voltage drop across the ith element in the loop.

11.33 What is the dual of the eighth supplemental discovery?

THE MAGNETIC DUAL OF THE EIGHTH SUPPLEMENTAL DISCOVERY: *Magnetomotive force can be neither generated nor destroyed in a conservative magnetic field.*

11.34 What is the magnetic Kirchhoff's voltage law?

MAGNETIC KIRCHHOFF'S VOLTAGE LAW (MKVL): *In any magnetic network and at each instant of time, the algebraic sum of the magnetic voltages (F) around a closed loop must be equal to zero, that is,*

$$\sum_{i=1}^{M} F_i = 0 \tag{11.22}$$

Where M is the total number of branches in the loop and F_i is the branch magnetic voltage in the ith branch.

We can write this equation more explicitly as

$$\sum_{k=1}^{M_1} mmf_k = \sum_{i=1}^{M_2} F_i \tag{11.23}$$

CHAPTER 12
Answers

QUIZ 12.1

12.1 What is the role of postulates in a science?

It is interesting to note that even though postulates of physics are a cut-down and are actually different than the original postulates (or the primary postulates) that have gone into building the physical universe, it nevertheless exists as a workable subject within a certain range.

12.2 What is the one element, which should never be forgotten while studying a science?

The one essential element that should never be forgotten in studying physics and engineering is the use of "viewpoint" (postulate zero) that has gone into the construction of theories of classical or quantum physics. This viewpoint is a composite viewpoint, which is evolved and polished through millennia of thinking men on the subject. These are the valuable final product of practical philosophers and scientists, who by using the scientific methodology as their stock in trade, observed the material universe (the tangible) and its many natural phenomena and through trial and error and honed their way to superior theory and eventual knowledge (the intangible).

12.3 Draw a diagram showing the inter-relation between different postulates leading to the successful evolution of the classical and quantum physics.

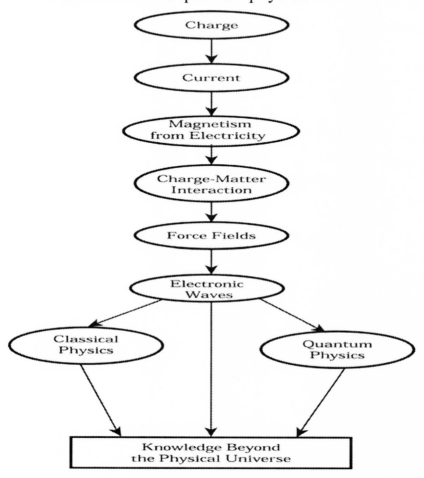

12.4 What lies ultimately beyond the physical entity of a product? What are the intervening steps?

Conversely, one can follow this line of logic and observe that before an application mass can exist there must be a manufacturer, before that a designer, and before that a postulator of that application mass; and before all of these, there comes a viewpoint that is existing exterior to this process and is solely concerned with observation and postulation of new and orderly thoughts, which reflects nature and its inner workings.

12.5 What must have existed before there was mechanical space, created energy, matter, and mechanical time?

So one can see that beyond the physical universe lies thought in its purest form: an "analytical and aware thought" that has the ability to create postulates, form concepts create space, create energy, matter and put them in motion as an application mass and thus create a time stream.

QUIZ 12.2

12.6 On a macro-scale, what are the main considerations that led to the construction of the solar system?

CONSIDERATION #1: *Let a space spanning approximately 8 billion ($8x10^9$) light years be created and setup to be called the physical universe. This is an approximate estimate and not an exact number representing the size of the physical universe.*

CONSIDERATION #2: *Let billions of islands of matter called galaxies of various sizes and shapes (e.g. Normal Spiral, elliptical, Barred Spiral, etc.) be created, each containing billions of stars, cloudlike patches of matter, clouds of hydrogen and dust be created (Note: the word galaxy comes from the Greek word "gala," which means "milk").*

In an aggregation of twenty galaxies known as "The Local Group", there is an elliptical pinwheel of 100 billion stars (where each star is similar to our sun), called "The Milky Way Galaxy". Our galaxy's shape is relatively two-

dimensional (i.e., flat) and is similar to a gigantic lens or a thick watch. Its dimensions are approximately 100,000 light years (or $3x10^{17}$ miles) with a maximum thickness of 10,000 light years. Let the entire Milky Way galaxy turn about its center at a speed of 150 miles a second (or 540,000 miles per hour). Because of its tremendous size, it takes the Milky way two hundred million years to make one complete turn.

CONSIDERATION #3: *Let the solar system be located on one of the spiral arms of the Milky Way galaxy, about 33,000 light years from the center of the galaxy and 17,000 light years from its edge. The galaxy is turning on an axis through the center, thus the solar system circles around the center of galaxy once every 200 millions of years. This turning motion is in addition to the motion of the galaxy as a whole through space relative to other galaxies.*

NOTE: *A light-year (ly) is the distance that light travels in one year, knowing well that the distance light travels in one second is 186,000 miles. Therefore one light-year would correspond to $9.461x10^{12}$ kilometers or $5.879x10^{12}$ miles. For example, the distance from the sun to the earth is approximately 8.3 light-minutes.*

CONSIDERATION #4: *Let the solar system be 3.5 billion miles (or 5.227 light-hours, where a light-hour (lh) is the distance that light would travel in one hour) in size consisting of a sun and a family of celestial bodies. The mass of the sun is 700 times more than the total mass of all attendant bodies together. The sun forms a powerful and massive center which holds all of the attendant celestial bodies in orbit due to its power of gravitation, very much like a merry-go-round machine in action.*

CONSIDERATION #5: *Let nine main planets revolve around the sun, in only one preferred direction. For example earth* *travels at 66,600 miles per hour on its orbit around the sun in a counterclockwise fashion when the observer is exterior to the solar system and views it from the celestial North Pole.*

CONSIDERATION #6: *Let planets as they revolve also spin around on an axis, which is at a slight tilt with their path of revolution around the sun. Planets spin in a preferred direction, which is a fixed direction due to the original postulate(s). In addition to the main nine planets, there are a thousand or more asteroids and other minor planets. For example the earth spins about itself counterclockwise when looking down on north pole (this is the preferred direction and the reason why the sun rises in the east and sets on the west) at a constant speed of approximately 1000 miles per hour that corresponds to one revolution per approximately twenty four hours (exactly 23 hours, 56 minutes and 4.1 seconds). Its tilt of 23.5 degrees provides a precession, which leads to change of seasons and variation in the length of day and night.*

CONSIDERATION #7: *Let the planets be at a certain distance from the sun and revolve around it in a particular orbit at an exact and constant speed. The orbits are not necessarily circular and the sun is not exactly at the center of the orbits. For example earth's orbit around the sun is 595 million miles long.*

CONSIDERATION #8: *Let all planets lie on an almost two-dimensional plane. This will be like a wheel where the sun lies at the hub and all other celestial bodies are distributed in the flat plane extending between the hub and the outer rim of the wheel.*

CONSIDERATION #9: *Let each planet have an exact diameter with a certain density and composition which leads to a certain surface gravity. For example, the* *diameter of earth is about 7900 miles from the North Pole to the South Pole, whereas the diameter at the equator is larger by 27 miles, which means earth is not exactly a sphere but flattened at the poles (The total perimeter of earth is about 25,000 miles). Furthermore, the South Pole is larger than the North Pole which makes earth shaped slightly like a pear. Its weight is approximately $6.6x10^{24}$ kg. Descending inward from the surface toward the center of the earth, we find that it consists of a series of concentric shells, three of the main ones are: a) a shell of rocks, b) an underlying granite shell, and c) a core of nickel and iron. Moreover, there are transitional layers between each of these main three shells.*

CONSIDERATION #10: Let there be four inner dwarf planets (i.e., Mercury, Venus, Earth and Mars) followed by four giant planets (i.e., Juniper, Saturn, Uranus and Neptune) and end with a dwarf one (i.e., Pluto) at the farthest outer rim.

CONSIDERATION #11: Let the sun consist of hydrogen which converts to Helium through a nuclear conversion thus creating an intense and continuous source of light and heat with a surface temperature of 11000 °F. The sun is the main source of radiated energy for all the celestial bodies in the solar system.

CONSIDERATION #12: Let some of the planets gather around them a total number of thirty one moons and satellites. For example, let earth have only one moon that orbits it at a fixed radial distance and at a certain uniform speed.

12.7 On a micro-scale, what are the main considerations that have gone into the structure of an atom?

CONSIDERATION #1: *Let a number of uncharged elementary particles (called neutrons) having a mass (m=1.67264x10^{-27} kg) be brought into close proximity. Bringing these particles together requires work to be done since these particles have a gravitational field. Thus this process creates a stored energy due to proximity of particles, which we call gravitational potential energy.*

CONSIDERATION #2: *Let a number of positively charged subatomic particles (called protons), each with a charge of Q$_p$= +1.60218x10^{-19} C and a rest mass of m$_p$=1.67264x10^{-27} kg (same as neutron's mass) be brought into close proximity with each other as well as to the group of neutron particles as gathered in the previous step. The neutrons and the protons as a whole form the center of an atom which is usually referred to as the nucleus of the atom.*

CONSIDERATION #3: *Let a number of charged subatomic particles (called electrons) each with a negative charge of Q$_e$= -1.60218x10^{-19} C and a rest mass of m$_e$=9.1095x10^{-31} kg (same as proton's charge but with a much smaller mass of approximately 1/1836 of a proton's) be brought at certain orbits around the nucleus. Obviously the mass of the atom is at the center with its huge gravitational pull, which prevents the electron to fly away under equilibrium conditions.*

CONSIDERATION #4: *Let the electrons revolve around the nucleus at fixed and unalterable orbits for each atom of the same material. The electron's position from the nucleus causes the atom to have a built-in potential energy due to proximity of the positive and negative charges. Furthermore, there is a gravitational built-in potential energy due to the proximity of masses of the three types of particles involved. The electron's motion in the orbits,* *being at well-defined speeds (determined through quantum physics with a certain probability), indicates a built-in kinetic energy in the atom as well.*

CONSIDERATION #5: *Let there be two electrons with opposite spins in each orbit. Each orbiting electron creates a magnetic field, which is perpendicular to the plane of the orbit (see Chapter 6, "Electricity from magnetism"). However in most cases, the two orbiting electrons with opposite spins cancel each other out, magnetically speaking!*

CONSIDERATION #6: *Let the total number of electrons be equal to the number of protons so that each atom is electrically neutral with no net charge.*

CONSIDERATION #7: *Let the number of electrons, protons and neutrons create different basic elements of matter, each with different and peculiar properties, such as Hydrogen, Oxygen, Iron, so on and so forth.*

CONSIDERATION #8: *Let the number of electrons in the outer orbit dictate the type of bonds that will be created when combined with other atoms.*

CONSIDERATION #9: *Let the atoms have the property of combining with the same or other atoms to form groups of atoms (such as molecules, etc.) to bring forth innumerable combinations of atoms and a host of materials found either in nature or generated synthetically. There are three main types of atomic bonding: a) Ionic bonding (abduction of electrons causes ions to form and attract each other), b) Metallic bonding (metal ions in a sea of electrons), and c) Covalent bonding (atoms sharing electrons).*

CONSIDERATION #10: *Let any and all of the physical universe materials (such as solids, fluids, etc.) and objects be composed of atoms and molecules so created and be only based upon these set of postulates and no other* *considerations. Figure 12.9 shows an atom as a building block, from which any material form (such as a crystal lattice) can be obtained.*

12.8 On a mini-scale, what are the main considerations that have gone into the design and construction of an integrated circuit?

CONSIDERATION #1*: Let the size of the chip be such that it creates a viable number of chips from the wafer. For example for a three inch wafer the minimum size of the chip to break even, for a yield of 100%, is 7.7 mm (see Technical References, #59, Chapter 21).*

CONSIDERATION #2*: Let the type of transmission lines and devices be chosen such that it will be integrable when built, with other elements on the circuit board*

CONSIDERATION #3*: Let the substrate material be such that it can accommodate the desired frequency range. For example, at low frequencies we use Silicon, whereas we need to use Gallium Arsenide (GaAs) or Indium Phosphide (InP) for high microwave and millimeter-wave frequency applications.*

CONSIDERATION #4*: Let the chosen device (bipolar or unipolar) be able to handle the maximum power consumption requirements and deliver the desired minimum output power.*

CONSIDERATION #5*: Depending on the application, the integrated circuit should be manufactured as a hybrid or monolithic IC. For example for mass production, and highly reliable, repeatable and broadband performance we need to use monolithic ICs, whereas for custom-made circuits we use hybrid ICs.*

CONSIDERATION #6*: Let the dielectric and conductor materials for the fabrication of the transmission lines,* *resistors, capacitors, etc., on the chip be selected based on the design requirements. For example for low resistance and highly conductive lines we need to use gold, while for high temperature applications we employ beryllia for the substrate.*

CONSIDERATION #7: *This consideration is along the lines of "what type of functions do we have to cram into the chip in order to be able to stay competitive in the market?", so on and so forth.*

12.9 How can one's life be a universe? Explain.

As a side comment and an analogy to what we have been discussing, we can observe that one's own life appears to be a universe of created things.

12.10 What is the personalized application mass (P.A.M.) of one's life?

The viewpoint furnished by the person himself is constantly in a state of postulation of future efforts, future goals and future personalized application mass (P.A.M.), which we may call personal possessions, for short. In a similar fashion, we can see that one's physical possessions, or the physical body and its characteristics (such as its shape, hairdo, etc.), where the body can be considered to be analogous to a machine, are actually the personalized application mass of the viewpoint, furnished by the life force.

12.11 How do postulates set up one's space?

Furthermore, one can see that since postulates make the space of a universe and since one's postulates shift in time, then one's space changes as time progresses.

QUIZ 12.3

12.12 Francis Bacon, as a champion in the rebirth of scientific thinking, developed the main goal of a science, what was it and what did it mean?

The Goal of sciences is stated as: "The sure and lawful goal of science is none other than this: That human life be endowed with new discoveries and power".

What actually Francis Bacon was referring to was that a science should give more enlightenment, more understanding and more power of choice to an individual over his environment and allow him to grow into a thinking and powerful being that he is; rather than turning him into an unthinking, automatic and stimulus-response machine who is fettered by old ideas handed down by religious authorities.

12.13 Describe how the physical sciences were considered to be a revolt against orthodox and entrenched religious authorities of the time?

Based on the antiquated and yet strong ideas which were dominant in the society at that time, the idea of science, as an independent thinking process, was a revolt against orthodox religion, which was entrenched as authority at the time and was heavily controlling the thoughts and the minds of the individuals in the society.

12.14 Is the science of physics truly a viewpoint-less science as it has been portrayed to be? Why?

No! It is definitely viewpoint dependent because:

The whole work was an attempt to understand the physical universe free from any bias or vested interest. This effort was purely from the viewpoint of the physical universe. Therefore, the pioneering scientists and philosophers of this era could be said, to have developed the science of physics by adoption of "The Physical Universe's Own Viewpoint", so to speak. They carefully avoided any discussions about the source or the postulator of the physical universe. This was rightly so, because such a discussion with the religious philosophers or scholars of the time would not have yielded any workable postulate or any scientific or unbiased truth worthy of considerations. Hence, the unknowing adoption of the physical universe's viewpoint, totally unspoken in any physics textbook is what we have inherited.

12.15 What was Einstein's realization about physics that led to the theory of relativity?

A few centuries had gone by and physics had remained stagnant and dead-ended in its tracks. This is where Einstein, in 1905, suddenly realized that physics, as workable as it was at the physical universe level, had no viewpoint associated with it. Therefore, he hurriedly tried to implement this missing vital element back into the subject. He introduced the concept of inertial frames of reference in his "Special theory of relativity," circa 1905. Approximately ten years later (circa 1915), he generalized this concept and proposed the "General theory of relativity" by introducing the non-inertial frames of reference.

12.16 What is the difference between a frame of reference and a viewpoint?

The substitution of frames of reference for the concept of "viewpoint—the cause point" is a very poor substitute and does not do justice. By frames of reference, Einstein meant purely an observer and not a viewpoint or a postulator of thought. These are totally different concepts.

12.17 What viewpoint (if any) does physics take on?

Therefore, the pioneering scientists and philosophers of this era could be said, to have developed the science of physics by adoption of "The Physical Universe's Own Viewpoint", so to speak.

12.18 Why is physics a study of total effect?

Today, physics remains purely a study of the physical universe not from the standpoint of "viewpoint" or "postulator" of space, energy, mass and time, but totally from the viewpoint of an "observer" of "created spaces, created energies, created masses" and the ensuing automatically "created time". Therefore, it can be said at this time that physics is a study of "total effect" and is not interested in finding the actual cause(s), but only the "apparent cause(s)" on a limited basis. It has made no attempt to relate its findings to other sources of life or energy that is not resident in the physical universe. In this regard, it can be said that it has remained a rather stagnant subject—on a second thought, maybe the founders of physics went down this path quite knowingly and perhaps wanted it to be viewpoint-less from the inception, even though we know well that it is impossible to be totally viewpoint-less! So the very confusing state of affairs in physics goes on!

QUIZ 12.4

12.19 What is the collective viewpoint of physics?

This viewpoint accumulated through thousands of years has shifted considerably through time and currently has set itself up to assume the viewpoint that views the space of the whole physical universe and has the task of understanding it from the atomic level to planetary, to galaxy level and beyond. It has several divisions such as elementary physics, classical physics, quantum physics, particle physics, astro-physics, etc.

12.20 What is the keynote of classical physics?

From the viewpoint of classical physics' practitioners, everything is certain and can be precisely analyzed or determined with a very high degree of accuracy.

12.21 What is the keynote of quantum physics?

From the viewpoint of quantum physicist, everything is uncertain and only a probabilistic analysis would show the probability of the occurrence or existence of any event. This is because the particles and energies within the created space only exist on a probability level, therefore all of the events or actions concerning these can only be expressed as a probability.

12.22 Why do concepts of classical physics clash with that of quantum physics?

From the viewpoint of quantum physicist, everything is uncertain and only a probabilistic analysis would show the probability of the occurrence or existence of any event. This is because the particles and energies within the created space only exist on a probability level, therefore all of the events or actions concerning these can only be expressed as a probability.

12.23 What is meant by an apparency? Give an example.

By apparency we mean a phenomenon or a condition that appears or seems to be a certain way, which is not the same and really different from its actuality. For example a spoon in a glass of water appears to be broken at the interface. This is an apparency, which differs from the actuality of the situation. The actuality is that rays of light undergo refraction at the air-water interface and slow down as they enter the water. This causes the rays to bend at the interface and thus make the spoon appear to be broken.

12.24 What are the postulates of quantum physics?

1) *Existence of a microscopically small "created space", which is uncertain.*
2) *Existence of fast-moving particles, which are difficult to trace but are associated with a probability of existence, or a probability wave packet.*
3) *Existence of the Heisenberg uncertainty principle.*
4) *Conservation of energy-mass system as a whole where mass and energy are inseparable in terms of nature of origin but convertible.*
5) *All calculations and results are to be expressed by a probability value and never with 100% certainty.*

12.25   Why does the physical universe appear to be infinite? How can one make it achieve its actual and finite size?

One is apt to assume an infiniteness about the physical universe, which is a total apparency and not an actuality. This apparency is caused by a seeming un-boundedness of the physical universe, which projects the ideas that no matter where we go, we can not get out of the physical universe. Thus the key to the size of the physical universe lies in one's ability to get exterior to it, at which moment the physical universe becomes a finite universe.

12.26 What is the generalized concept of a "viewpoint"?
However, the role of a viewpoint in the creation of a whole planet, a solar system, a galaxy or in general the whole physical universe may not, at first glance, be so evident. This is where we need to resort to inductive logic and based on the evidence on a smaller scale, assume or better yet postulate the existence of a viewpoint (or collective viewpoints) on a larger scale, which has caused the creation of what we call the whole physical universe.

QUIZ 12.5

12.27 What is the first visible thing of any universe?
The application mass.

12.28   What are the steps one needs to do in order to improve an existing universe or make up a whole new universe?

   a) *One needs to understand the basic postulates or considerations that have been utilized in the construction of any universe under study (such* as the physical universe, one's own universe, etc.),

b) *One needs to create or make new postulates or considerations such that they either generalize or encompass the old ones, which then allow a basic change to occur in the universe one is studying, and*

c) *Finally, based on the new postulates, one derives all of the related axioms, laws and design techniques and builds a whole gamut of new application mass to replace the old ones. This process would surely bring about a whole new universe or field of study and can be called "creating a new universe".*

12.29 How did Aristotle help create the modern field of "digital logic"?

Over 2000 years ago the Greek philosopher Aristotle postulated the "Digital Universe", which is based upon the two-valued logic system (the true or false system) consisting of three main postulates. With this, he ushered in and thus established a whole new field of reasoning and study, which we now call "digital logic". Over a period of two thousand years we have gone through quite a bit of evolution and gradually but finally have reached to its associated application mass in terms of numerous digital devices, circuits, components, computers, and systems.

12.30 What lies between the postulates and a viewpoint?

Purposes; a series of postulates are made solely based upon a series of purposes. The purposes dictate the extent and shape of the postulates.

12.31 What does the physical universe constantly invite us to do?

The physical universe is a created universe, which invites us almost on a constant basis to explore in order to discover its very many intricacies or facets of existence: on a microscopic scale (e.g., new subatomic particles, atoms, molecules, etc.) or on a macroscopic level (new phenomenon, planets, stars, galaxies, etc.).

12.32 Can one derive all of the axioms, natural laws, rules and theorems from a series of postulates? Explain.

Of course, it goes without saying that once we have discovered the actual postulates concerning any portion of the physical universe, it is only a matter of time before one derives or discovers a series of governing and irrefutable laws that will completely describe the subject matter fully, and will delineate the behavior of all of the related particles, elements, objects, etc., that would supplement the postulate portion.

12.33 How can the physical universe be classified as a "total effect" type universe? Explain.

The physical universe happens to be a "consequence-happy" universe, which means that for every action one may undertake in it, one should be prepared to receive a consequent reaction, which as a result has turned it into a dangerous and harsh universe. It does not care to understand but only wants to be understood. For example a cup falls off of a table; it receives instantly a reaction and shatters without any thought from the physical universe. To put it mildly, it is an "unthinking universe of force". It has no concourse with reason or logic.

12.34 Why is there a relentless effort to understand the physical universe?

We as human beings are on this relentless quest to analyze and unravel the many "secrets" of this created universe without ever stopping for a moment and visualizing the bigger scheme of affairs, which includes the realization of the fact that before there was an electron or an atom there was the thought of an electron or an atom. Before there were these massive galaxies, stars and planets, there were the thoughts of galaxies, stars and planets. Thus the actual things to explore are the considerations or the thoughts behind our created universe, and not the created objects themselves endlessly.

12.35 What is meant by the universe ahead and how will it be from the point of view of sciences?

On the other hand, "the universe ahead", which is constantly being shaped or carved out of the current physical universe by the beings who occupy it, will be a far more "dynamic and gentler" universe. We have and are surely developing a new universe with the force of our reasoning, a force which this universe has no concourse with and has never seen before.

With every revolution of earth we take one more step in the direction of creating this universe. It will be much superior to the present one, which we have been observing for the last 100 thousand years as a species of human beings.

QUIZ 12.6

12.36 What are the three categories of application mass? Explain.

There are three categories of application mass:
a) Generalized Application Mass (G.A.M.)
b) Technical Application Mass (T.A.M.)
c) Personalized Application Mass (P.A.M.)

12.37 What is the difference between G.A.M. and T.A.M.?
Technical Application Mass (T.A.M.) is the category of man-made application mass that is produced directly as a result of application of a science using its scientific postulates, axioms, laws and other technical data, whereas Generalized Application Mass (G.A.M.) is the very general category of all masses (liquids, solids and gases) found in nature and existing in a raw and unprocessed form.

12.38 What is the totality of all of the application masses (G.A.M, T.A.M. and P.A.M.) called? Explain.
The totality of all types of application mass (i.e., G.A.M., T.A.M. and P.A.M.) is referred to as "mechanics," which is commonly used amongst philosophers and scientists to refer to anything not imbued with life force. Since mechanics, in general, includes particles ranging in size from an electron to an object or a projectile (such as a missile, rockets, etc.) thus its study perforce requires the use of quantum mechanics at the microscopic scale and the laws of classical mechanics at the macroscopic level. For the study of super-sized particles such as planets or galaxies, we need to utilize the laws of astro-physics for proper interpretation of the observed data.

12.39 What is the link between sciences and philosophy?

The three categories of application mass give us a panoramic view of the visible world, with which we find ourselves in direct contact everyday. Furthermore, this panoramic view of application mass is equivalent to what philosophers normally refer to as "The Mechanics," which is the mechanical aspects of life and existence. Thus we have connected the "physical sciences" to a much larger sphere of knowledge called "philosophy" through the use of the pyramid of knowledge (see Chapter 1), and actually have made them one of its many subsets.

12.40 What is nano-technology and what impact will it have on our lives when it fully matures?

Nano-technology deals with the microscopic universe and therefore could be considered to be a subset of quantum mechanics. It is the science of working at atomic dimensions to engineer materials and machines out of individual molecules, and is based upon the same set of postulates and axioms that we have discussed in Chapters 1, 2 and 3.

12.41 What are the essential postulates and why knowing them is important?

There are six essential postulates that form the most fundamental aspects of electricity. These six essential postulates state the most distilled concepts about electricity, from which we can derive all of the major laws of electricity, magnetism, electromagnetism and electromagnetic optics.

12.42 What are the auxiliary postulates?

There are also several auxiliary postulates that are not as basic but help to understand electricity better. From these six auxiliary postulates, some of the important laws of electricity can be derived.

12.43 What does the future of the physical universe depend on?

The physical universe, as a total-effect type of universe, being a total byproduct of the original postulates, has a future which is proportional to our complete understanding of its postulates and its many derived features. The upcoming physical universe is a direct function of how well we understand the postulates clearly laid out in this book as well as our own relationship with regard to it and the role that our thoughts play in its existence. Furthermore, how successfully we tame its many savage forces and put them to use toward enhancing our own survival, while harnessing our own destructive impulses in order to prevent its ultimate demise, will have a huge bearing on the fate of the earth, the planet we call our home at the moment in this forlorn corner of the universe!

CHAPTER 13
Answers

QUIZ 13.1

13.1 What is the primary factor interwoven in all aspects of the physical universe? Explain.

The primary factor intertwined in all aspects of the physical universe is considerations, considerations and more considerations. The number of considerations built into the physical universe approach infinity in sheer numbers. These considerations appear invisible at first glance but reign in all aspects of the physical universe. Every known scientific fact about the physical universe is actually a consideration and cataloguing them in a list format would fill many volumes of books.

13.2 What is any portion of the physical universe constructed from?

Descending from the level of considerations, the whole physical universe or any part or sub-region thereof consists of a created space with a certain amount of energy placed in it.

13.3 What are the main two subdivisions of energy particles? Give examples.

The energy can be subdivided into light particles (such as photons and electrons) and heavy particles forming particles of matter (such as atoms and molecules). Particles of matter are obtained by a condensation of lighter energy particles.

13.4 How can we consider matter to be a standing wave?

The whole physical universe with all of its gigantic galaxies, billions of stars and billions more planets can be considered as standing waves floating on a relatively timeless basis (i.e., forever), existing in a space, which was created earlier in time than energy or matter. The time frame of existence of these galaxies and planets is on the order of trillions of years, which vastly exceeds man's length of existence as a species on earth and his continued survival since his appearance less than 100 thousand years ago.

13.5 What makes matter and energy to be related to each other?

Since the temperature is always above absolute zero, therefore energy in the physical universe is in constant motion and has a wave property associated with it. Furthermore, matter can be considered to be a ridge of energy or a standing wave, which is floating in time. Moving matter also has a wave associated with it according to the de Broglie's theory (see Chapter 3, section 3.7, Particle and Wave Duality). Therefore, we can conclude that since all of the energy and matter in the physical universe is in constant motion, thus we are dealing with waves as the root and source of the physical universe. The motion of waves brings forth a time stream.

13.6 What are the two main facets of the physical universe?

In actuality, we are dealing with a universe, which is two-faceted: a) One facet existing at a microscopic level, which is hidden and yet with extreme dominance underlies and actually makes up everything, and b) The second facet existing on a macroscopic scale, which brings about a level of reality with which we are too familiar and deal with on a daily basis

QUIZ 13.2

13.7 What is the microscopic universe? What is another name for it? Explain.

THE MICROSCOPIC UNIVERSE: *This universe is the "actuality version" (also called the "underlying-theme version") of the physical universe, and consists of a series of light or heavy energy particles (photons, electrons, atoms, etc.) which are moving so fast that determination of their speed, energy or position with a high degree of certainty is impossible,*

13.8 What is the macroscopic universe? Why is it called the "apparency version" of the physical universe?

THE MACROSCOPIC UNIVERSE: *This universe is the "apparency-version" (also called the "visible version") of the physical universe and consists of a series of many particles grouped together where each group of particles is formed into one macro-particle and acts as one coherent unit. These macro-particles are usually referred to as objects or bodies (such as baseballs, cars, airplanes, planets, galaxies, etc). The objects at this level of observation follow certain and definite laws of action or motion, which behaviorally are very predictable and mathematically, very precise.*

13.9 Why does the microscopic universe dominate the macroscopic universe and every aspect of our lives on a physical plane?

At the macro-particle level, we have an apparency, where it seems that the macroscopic version of the physical universe is understandable and appears to bring forth a high level of certainty and predictability with it. However, since the microscopic version of the physical universe is the underlying theme of all things, therefore it is the most dominant and overriding factor in all of this.

13.10 What is the real nature of the physical universe?
Since the microscopic universe actually feeds the entire macroscopic universe, thus any and all of the observed motions and actions at the macroscopic level become glazed and laced with a layer of uncertainty. This is where the axiom of "absolutes are unachievable in the physical universe but only approachable" takes its cue from, as discussed in Chapter one.

13.11 What factor makes the physical universe appear to be orderly and organized?
The beings in it make it seem orderly and organized to a very small degree but truth be told this is a universe, which keeps itself hidden by constantly moving at a rapid rate and not telling us where it is going to. It asks one constantly to understand its confusions and find out its secrets.

QUIZ 13.3
13.12 What are the three original postulates?
There are three main governing postulates in the physical universe that has a total effect on every aspect of it. These three are referred to as the "Original postulates," which

were made very early, at the blue print stage, of the physical universe, as follows:

A) ORIGINAL POSTULATE #1- EXISTENCE OF SPACE: *This is the first postulate made very early, at the beginning of the physical universe.*

B) ORIGINAL POSTULATE #2- EXISTENCE OF ENERGY (or Force): *This is the second postulate made after the first postulate of the physical universe and gave birth to the first particle of energy and many more soon to follow.*

C) ORIGINAL POSTULATE #3- EXISTENCE OF CHANGE: *This third postulate followed right after making the above two original postulates and gave birth to the first particle of matter and the very first few seconds of mechanical time many trillions of years ago.*

13.13 What do we know about the mind of "God"? Explain.

The three original postulates, even though considered to exist sequentially, (i.e., first space then energy and finally change), however, when it comes to their actual implementation they must be simultaneously executed. Therefore, we can see that by understanding these postulates, we have taken a great stride in the direction of knowing the mind of God!

13.14 What are the salient points into which any subject can be boiled down?

In the final analysis, we can boil down all of our discussion to only a few salient points:

a) *There is first a viewpoint (furnished by the life force), which is central to any universe. The viewpoint makes up several purposes, each leading to several (or many) postulates,*

b) *The postulates, as set forth by the viewpoint and derived from the purposes, create universes, where one of these universes, when shared with all other viewpoints, is called the physical universe,*

c) *From this commonly-shared universe (i.e., the physical universe) and its associated postulates, we can extract all of the scientific laws and axioms,*

d) *Using these exact laws and axioms and with the help of a series of theorems and an abstract language such as mathematics, a number of design rules and manufacturing principles can be obtained, from which a plethora of technical application mass is created,*

e) *Finally, other viewpoints (such as observing viewpoints), view the "Generalized Application Mass" and may wonder about the incredulity of its existence or the enormity of its creation and thus try to understand them through scientific research or otherwise.*

13.15 What makes the subject of postulates an extremely important subject when dealing with the construction of any universe? Explain.

Knowing the mind of God is associated with knowing the subject of postulates, which is an interesting and extremely vital subject, a subject that is worth studying since its full understanding can open doors to new universes, many unheard-of vistas or soon-to-be-created universes. It is a powerful subject and its study is vastly more rewarding than any rigorous education in pure physics or any other extant physical science.

QUIZ 13.4

13.16 What is the one single thing that makes or breaks any given universe? Why?

We should understand the power of postulates and the impact they have upon every aspect of our own internal universe as well as "the shared universe" or "the physical universe". Furthermore, we need to realize that one's postulates can gain command value over the physical universe if they are well-conceived, precisely laid out and properly executed; thus one can do quite a bit of postulating, which will create a better technical world benefiting self and others.

13.17 What lies beyond postulates? Explain.

Progressively, Purposes, Viewpoint, and finally the Life Force, in the reverse order of importance

13.18 Draw a pyramid of knowledge for the field of electricity and write down all of the important layers in the pyramid.

The diagram is shown on the next page.

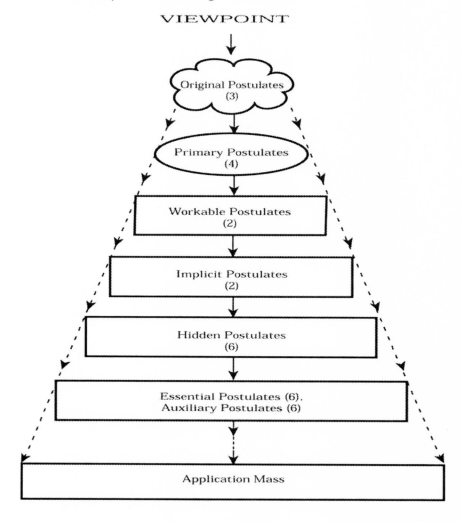

13.19 What are the two main subdivisions of electricity and what are the ramifications for each?
a) Static electricity, leading to charge storage effects, and
b) Kinetic Electricity, leading to current flow, magnetism, and wave phenomena.

Appendix C

ANSWERS TO THE FINAL QUIZ

The Final Quiz
Answers

F.1 Using the book, what are the most important considerations used in the construction of our material universe?

Three original postulates leading to four primary postulates, which are described as:

I. ORIGINAL POSTULATES: These three postulates cover the most irreducible set of postulates from which the entire physical universe can be derived and constructed.

II. PRIMARY POSTULATES: These four postulates, in simultaneous implementation, bring the whole material universe into existence and actually present a more tangible and understandable version of the original postulates. Therefore, primary postulates deal with the fundamental considerations concerning the physical universe.

F.2 What is a universe in general and what are the elements that comprise it?

A universe is a system of created things and has application mass as the most visible part of it with the postulates and viewpoint as most invisible part of it.

F.3 In the field of electricity, what are the most essential concepts that have led to its progress and its current development?

The Six monumental discoveries are:

a) THE MONUMENTAL DISCOVERY #1: *Realization of Static Charge*

1) William Gilbert
2) Charles A. Coulomb

b) THE MONUMENTAL DISCOVERY #2: *Realization of Kinetic Perpetual Charge.*

a) Luigi Galvani
b) Alessandro Volta

c) THE MONUMENTAL DISCOVERY #3: *Realization of Magnetism from electricity.*

Hans Christian Oersted

d) THE MONUMENTAL DISCOVERY #4: *Realization of Kinetic Charge-Matter Interaction.*

Andre-Marie Ampere

e) THE MONUMENTAL DISCOVERY #5: *Realization of Force Fields.*

Michael Faraday

f) THE MONUMENTAL DISCOVERY #6: *Realization of Electronic Waves.*

James Clerk Maxwell
Heinrich Hertz

F.4 Considering the subject of physical sciences, what led to their tremendous success and huge dominance in our world today?

The use of scientific methodology! Scientific methodology as given in section 1.8.3, is a systematic approach based on: a) observation, b) collection of data, d) classification of data, e) formulation of a hypothesis, e) testing and experimentation, f) repeat if refinement needed and g) emergence of a workable theory.

It is extremely superior to other methods because it delivers a workable answer that can be used over and over to solve any new problem and help us in predicting the future behavior of any complex system.

F.5 Based on your understanding, why social sciences have neither achieved the success nor the dominance that physical sciences have?

A lack of a series of exact postulates and inexorable axioms along with a no-understanding and a total neglect of the scientific methodology are the main reasons.

F.6 What is the one single concept that makes or changes the space of any universe? Explain.

The postulates put forth by the viewpoint (see section 1.3.2, The Mechanics of creation of any universe).

F.7 In your view, what methodology has led to making sciences the today's towering giant?

The scientific methodology.

F.8 How does the concept of dichotomy simplify understanding of sciences?

Because there is no one single thing to consider (i.e., no absolutes!) but also the opposite as well, which makes them a pair of data related together.

F.9 What are the four major steps used in any scientific investigation? Explain each step.

a) Observation and imagination,

b) Generation of a working hypothesis, followed by

c) *Experimentation to confirm the validity of the working hypothesis, and*

d) *Reiteration of this process until a valid hypothesis is obtained at which moment it becomes a principle.*

F.10 What is "basal unity" and what is meant by "the universal nature of force"?

Faraday had a philosophical approach to his work and believed in the "unity of physical phenomena", in particular, he considered that matter, electricity and light were of the same origin. He envisioned that a basal field of force (such as one exhibited by magnetism) accounted for the substance and operation of all nature. His vision was based on the oldest and yet continuing quest in the philosophy of science-- the Basal Unity in Nature i.e. all forces or energy sources come from a common source and therefore are convertible into one another.

F.11 How do sciences connect with the subject of philosophy?

Having established the three categories of application mass (T.A.M., G.A.M., and P.A.M.) we can see that regardless of its category, one still has to go through the very same process as described in Chapter one; that is starting with postulates at the top and move down to the base of the pyramid of knowledge to reach the desired application mass.

The above three categories of application mass give us a panoramic view of the visible world, with which we find ourselves in direct contact everyday. Furthermore, this panoramic view of application mass is equivalent to what philosophers normally refer to as "The Mechanics," which is the mechanical aspects of life and existence. Thus we have connected the "physical sciences" to a much larger sphere of knowledge called "philosophy" through the use of the pyramid of knowledge (see Chapter 1), and actually have made them one of its many subsets.

F.12 What are similarities and differences between the classical and quantum mechanics?

SIMILARITIES:
Space, matter, energy, and time.
DIFFERENCES:
The degree of certainty of observation, and probability of existence; which for classical physics is almost 100% and for quantum physics is a variable and is actually defined mostly by mathematical expressions.

F.13 What is the role of viewpoint in sciences?

The viewpoint is the dominant entity that controls any created universe.

F.14 What was on "God's mind," when he created the physical universe?

The three original postulates: space, energy and perpetual change.

F.15 What are the six field of study in electrical engineering? Describe each.

ELECTROSTATICS, Static electric field or electricity,
ELECTROKINETICS, Current flow,
MAGNETOSTATICS, Static magnetic field,
ELECTRODYNAMICS, Movement of matter by electricity,
ELECTROMAGNETICS, coexisting electricity and magnetic fields,

ELECTRONIC WAVES, propagating electric and magnetic fields.

F.16 What are the factors affecting the future of the physical universe?

The physical universe, as a total-effect type of universe, being a total byproduct of the original postulates, has a future which is proportional to our complete understanding of its postulates and its many derived features. The upcoming physical universe is a direct function of how well we understand the postulates clearly laid out in this book as well as our own relationship with regard to it and the role that our thoughts play in its existence. Furthermore, how successfully we tame its many savage forces and put them to use toward enhancing our own survival, while harnessing our own destructive impulses in order to prevent its ultimate demise, will have a huge bearing on the fate of the earth, the planet we call our home at the moment in this forlorn corner of the universe!

F.17 If you had to summarize this entire work into one paramount concept, what would that be? Why? Explain.

Understanding the true postulates of any science paves the road that leads to the realm of knowledge and enables one to gain the highly desired ability of critical thinking in that science. In other words, the neglected "subject of postulates" is the gateway to a higher level of understanding of the physical universe or the physical sciences; a subject, which is almost never mentioned or emphasized adequately by any physical scientist or technical educator as the superior factor. Moreover, the postulates as a whole furnish the metaphorical key that opens the door to the mysteries regarding the created things in one's own universe. It is the most powerful subject there is beyond the masses of the physical universe and its study is equally, or perhaps more, as rewarding!

Appendix D

CLASSICAL LAWS OF ELECTRICITY

(*Covering All Frequencies, from DC to Light*)

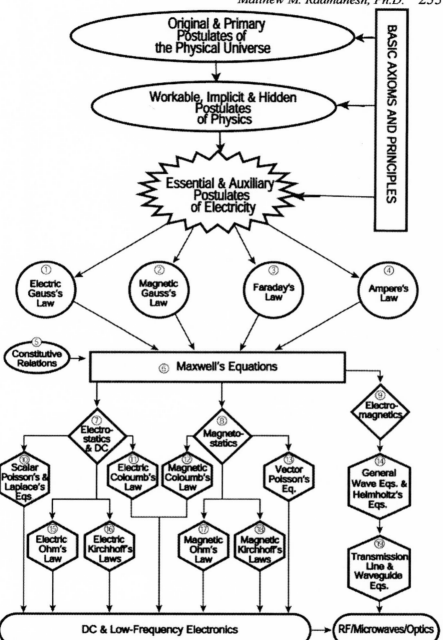

FIGURE D.1 All fundamental laws (and their duals) of
electricity applicable to all frequencies (DC to Light).

Glossary of Technical Terms

The following glossary supplements the presented materials in the text, but does not replace the use of an unabridged technical dictionary, which is a must for mastery of sciences.

Absolute
a) That which is without reference to anything else and thus not comparative or dependent upon external conditions for its existence (opposed to relative), b) That which is free from any limitations or restrictions and is thus unconditionally true at all times.

Absolute Temperature Scale
A scale with which temperatures are measured relative to absolute zero (the temperature of –273.15 °C or –459.67 °F or 0 K). The absolute temperature scale leads to the absolute temperatures, which are: a) The temperature in Celsius degrees, relative to –273.15 °C (giving rise to the Kelvin scale), and b) The temperature in Fahrenheit degrees, relative to –459.67 °F (giving rise to the Rankine scale). At absolute zero temperature, molecular motion theoretically vanishes and a body would have no heat energy. The absolute zero temperature is approachable but never attainable. *See also* **temperature**.

Active Device
An electronic component such as a transistor that can be used to produce amplification (or gain) in a circuit.

Admittance
The measure of ease of AC current flow in a circuit, the reciprocal of impedance expressed in Siemens (symbol Y or y).

Ampere (A)

The unit of electric current defined as the flow of one Coulomb of charge per second. Alternately, it can also be defined as the constant current that would produce a force of 2×10^{-7} Newton per meter of length in two straight parallel conductors of infinite length, and of negligible cross section, placed one meter apart in a vacuum.

Ampere's Law

Current (either conduction or displacement) flowing in a wire or in space generates a magnetic flux that encircles the wire in a clockwise direction when the current is moving away from the observer. The direction of the magnetic field follows the right hand rule (This law may also be referred to as the law of magnetic field generation).

Differential form: $\mathrm{Curl}\overline{H} = \overline{J} + \dfrac{\partial \overline{D}}{\partial t}$,

Integral form : $\oint_C \overline{H} \cdot \overline{d\ell} = I + \int_S \dfrac{\partial \overline{D}}{\partial t} \cdot \overline{dS}$

Amplitude

The extent to which an alternating current or pulsating current or voltage swings from zero or a mean value.

Analog

Pertaining to the general class of devices or circuits in which the output varies as a continuous function of the input.

Anode

The positive electrode of a device (such as a diode, etc.) toward which the electrons move during current flow.

Application Mass

All of the related masses that are connected and/or obtained as a result of the application of a science. This includes all physical devices, machines, experimental setups, and other physical materials that are directly or indirectly derived from and are a result of the application. In this book when we say application mass, we really mean "technical application mass." See also Generalized application mass, Technical application mass and personalized application mass.

Attenuation

The decrease in amplitude of a signal during its transmission from one point to another.

Attenuation Constant

The real component of the propagation constant.

Attenuator

A resistive network that provides reduction of the amplitude of an electrical signal without introducing appreciable phase or frequency distortion.

Average Power

The power averaged over one cycle.

Axiom

A self-evident truth accepted without proof.

B

Bias

The steady and constant current or voltage applied to an electrical device to establish an operating point for proper operation of the device.

Bidirectional

Responsive in both directions.

Bilateral

Having a voltage-current characteristic curve that is symmetrical with respect to the origin. If a positive voltage produces a positive current magnitude, then an equal negative voltage produces a negative current of the same magnitude.

Brewster Angle

The angle of incidence of light reflected from a dielectric surface at which the reflection coefficient becomes zero when the light's electrical field vector lies in the plane of incidence (parallel polarization). In other words, if a parallel polarized wave is incident at a dielectric surface at the Brewster angle, all of the wave will be transmitted through and there will be no reflection. Generally speaking, the concept of Brewster angle applies to any electronic wave of any frequency, not just light waves [$\theta_B=\tan^{-1}(\varepsilon_2/\varepsilon_1)^{1/2}$].

C

Capacitance

The property that permits the storage of electrically separated charges when a potential difference exists between two conductors. The capacitance of a capacitor is defined as the ratio between the electric charge of one electrode, and the difference in potential between the electrodes.

Capacitor

A device consisting essentially of two conducting surfaces separated by an insulating material (or a dielectric) such as air, paper, mica, etc., that can store electric charge.

Cathode

The portion or element of a two-terminal device that is the primary source of electrons during operation.

Cavity (Also Called a Cavity Resonator)

A metallic enclosure inside which resonant fields at microwave frequencies are excited in such a way that it becomes a source of electromagnetic oscillations frequencies.

Cell

A single and basic unit for producing electricity by electrochemical or biochemical action. For example, a battery consists of a series of connected cells.

Celsius (°C)

$1/100^{th}$ of the temperature difference between the freezing point of water (0°C) and the boiling point of water (100°C) on the Celsius temperature scale given by:

$$T(°C)=T(K)-273.15 =\frac{5}{9}\{T(°F)-32\}.$$

Characteristic Impedance

The driving-point impedance of a transmission line if it were of infinite length. This can also be defined as the ratio of the voltage to current at every point along a transmission line on which there are no standing waves. It is given in general by:

$$Z_0 = \sqrt{(R + j\omega L)/(G + j\omega C)}$$

Charge

A basic property of elementary particles of matter (electrons, protons, etc.) that is capable of creating a force field in its vicinity. The built-in force field is a result of stored electric energy.

Chip

A single substrate upon which all the active and passive circuit elements are fabricated using one or all of the semiconductor techniques of diffusion, passivation, masking, photoresist, epitaxial growth, etc.

Circuit

The interconnection of a number of devices in one or more closed paths to perform a desired electrical or electronic function.

Classical Mechanics (Also Called Classical Physics, Non-Quantized Physics or Continuum Physics)

Is the branch of physics based on concepts established before quantum physics, and includes materials in conformity with Newton's mechanics and Maxwell's electromagnetic theory.

Coaxial Transmission Line (Also Called Coaxial Cable)

A concentric transmission line in which one conductor completely surrounds the other, the two being separated by a continuous solid dielectric or by dielectric spacers. Such a line has no external field and is not susceptible to external fields.

Coulomb (C)

The unit of electric charge defined as the charge transported across a surface in one second by an electric current of one ampere. An electron has a charge of 1.602×10^{-19} Coulomb.

Coulomb's Laws

The laws that state that the force (F) of attraction or repulsion between two electric charges (or magnetic poles) is directly proportional to the product of the magnitude of charges, Q (or magnetic pole strengths, M), and is inversely proportional to the square of distance (d) between them; that is,

Electric: $F = \dfrac{Q_1 Q_2}{4\pi\varepsilon d^2}$,

Magnetic: $F = \dfrac{M_1 M_2}{4\pi\mu d^2}$

The force between unlike charges, Q_1 and Q_2 (or poles, M_1 and M_2) is an *attraction*, and between like charges (or poles) is a *repulsion*.

Communication Principle (Also Called Universal Communication Principle)
A fundamental concept in life and livingness that is intertwined throughout the entire field of sciences that states for communication to take place between two or more entities, three elements must be present: a source point, a receipt point, and an imposed space or distance between the two.

Complex Power
Power calculated based on the reactance of a component.

Component
A packaged functional unit consisting of one or more circuits made up of devices, which in turn may be part of an operating system or subsystem.

Conductivity
The ratio of the current density (J) to the electric field (E) in a material. It represents the ability to conduct or transmit electricity.

Conductor
a) A material that conducts electricity with ease, such as metals, electrolytes, and ionized gases; b) An individual metal wire in a cable, insulated or un-insulated.

Curl Operation
Curl is an operation on a vector field, which creates another vector whose magnitude measures the maximum net circulation per unit area of the vector field at any given point and has a direction perpendicular to the area, as the area size tends toward zero. The

cause of the curl of a vector field is a vortex source. For example electric current (conduction or displacement) is the vortex source for magnetic field.

Current

Net transfer of electrical charges across a surface per unit time, usually represented by (I) and measured in Ampere (A). Current density (J) is current per unit area.

D

DC (Also Called Direct Current)

A current which always flows in one direction (e.g., a current delivered by a battery).

Decibel (dB)

The logarithmic ratio of two powers or intensities or the logarithmic ratio of a power to a reference power, multiplied by 10. It is one-tenth of an international unit known as *Bel*: $N(dB)=10\log_{10}(P_2/P_1)$.

Device

A single discrete conventional electronic part such as a resistor, a transistor, etc.

Diamagnetics

are materials (such as glass, wood, lead, sulfur and others), which avoid magnetic lines of force.

Dichotomy

Two things or concepts that are sharply or distinguishably opposite to each other.

Die (Also Called Chip)

A single substrate on which all the active and passive elements of an electronic circuit have been fabricated. This is one portion taken from a wafer bearing many chips, but it is not ready for use until it is packaged and provided with terminals for connection to the outside world.

Dielectric

A material that is a non-conductor of electricity. It is characterized by a parameter called *dielectric constant* or *relative permittivity* (ε_r).

Dielectric Constant

The property of a dielectric defined as the ratio of the capacitance of a capacitor (filled with the given dielectric) to the capacitance filled with air as the dielectric, but otherwise identical in geometry.

Diffraction

Is the redistribution of intensity of waves in space, which results from the presence of an object (such as a grating, consisting of narrow slits or grooves) in the path of beam of lightwaves. This shall split up the beam into many rays, causing interference and thus producing patterns of dark and light bands downstream (i.e., regions with variations of wave amplitude and phase).

Digital

Circuitry in which data-carrying signals are restricted to either of two voltage levels.

Discovery

The gaining of knowledge about something previously unknown.

Discrete Device

An individual electrical component such as a resistor, capacitor, or transistor as opposed to an integrated circuit that consists of several discrete components.

Distributed Element

An element whose property is spread out over an electrically significant length or area of a circuit instead of being concentrated at one location or within a specific component.

Divergence

a) The emanation of many flows from a single point, or reversely, the convergence of many flows to one point; b) (of a vector field, F) The net outflux per unit volume at any given point in a vector field, as the volume size shrinks to zero (symbolized by divF). The cause of the divergence of a vector field is called a flow source. For example, positive electric charge is the flow source for the electric field and creates a net outflux of electric field per unit volume at any given point.

Dual
Two concepts, energy forms or physical things that are of comparable magnitudes but of opposite nature, thus becoming counterpart of each other.

Duality Theorem
States that when a theorem is true, it will remain true if each quantity and operation is replaced by its dual quantity and operation. In circuit theory, the dual quantities are "voltage and current" and "impedance and admittance." The dual operations are "series and parallel" and "meshes and nodes."

E

Electric Charge (or Charge)
(Microscopic) A basic property of elementary particles of matter (e.g., electron, protons, etc.) that is capable of creating a force field in its vicinity. This built-in force field is a result of stored electric energy. (Macroscopic) The charge of an object is the algebraic sum of the charges of its constituents (such as electrons, protons, etc.), and may be zero, a positive or a negative number.

Electric Current (or Current)
The net transfer of electric charges (Q) across a surface per unit time.

Electric Field
The region about a charged body capable of exerting force. The intensity of the electric field at any point is defined to be the force that would be exerted on a unit positive charge at that point.

Electric Field Intensity
The electric force on a stationary positive unit charge at a point in an electric field (also called *electric field strength*, *electric field vector*, and *electric vector*).

Electrical Noise (or Noise)
Any unwanted electrical disturbance or spurious signal. These unwanted signals are random in nature, and are generated either

internally in the electronic components or externally through impinging electromagnetic radiation.

Electricity
Is a form of energy, which can be subdivided into two major categories: a) Electrostatics, and b) Electrokinetics.

Electrodynamics
Is a scientific field of study dealing with the various phenomena of electricity in motion, including the interactions between current-carrying wires as well as the forces on current wires in an independent magnetic field.

Electrokinetics
Is that broad and general field of study dealing with electric charges in motion. It studies moving electric charges (such as electrons) in electric circuits and electrified particles (such as ions, etc.) in electric fields.

Electrolysis
The action whereby a current passing through a conductive solution (called an *electrolyte*) produces a chemical change in the solution and the electrodes.

Electrolyte
A substance that ionizes when dissolved in a solution. Electrolytes conduct electricity, and in batteries they are instrumental in producing electricity by chemical action.

Electrolytic Cell
In general, a cell containing an electrolyte and at least two electrodes. Examples include voltaic cells, electrolytic capacitors, and electrolytic resistors.

Electromagnetic (EM) Wave
A radiant energy flow produced by oscillation of an electric charge as the source of radiation. In free space and away from the source, EM rays of waves consist of vibrating electric and magnetic fields that move at the speed of light (in vacuum), and are at right angles to each other and to the direction of motion. EM waves propagate with no actual transport of matter, and grow weaker in amplitude as they travel farther in space. EM waves include radio, microwaves,

infrared, visible/ultraviolet light waves, X-ray, gamma rays, and cosmic rays.

Electromagnetics
The branch of physics that deals with the theory and application of electromagnetism.

Electromagnetism
a) Magnetism resulting from kinetic electricity; b) Electromagnetics.

Electron
A stable elementary particle of matter, which carries a negative electric charge of one electronic unit equal to q= -1.602x10^{-19} C and has a mass of about 9.11x10^{-31} kg and a spin of ½.

Electronics
The study, control, and application of the conduction of electricity through different media (e.g., semiconductors, conductors, gases, vacuum, etc.).

Electroplating
Depositing one metal on the surface of another by electrolytic action.

Electrostatics
The branch of physics concerned with static charges and charged objects at rest.

Elementary Particle
A particle, which can not be described as a compound of other particles and is thus one of the fundamental constituents of all matter (e.g. electron, proton, etc.).

Energy
The capacity or ability of a body to perform work. Energy of a body is either potential motion (called *potential energy*) or due to its actual motion (called *kinetic energy*).

F

Fahrenheit (°F)

$1/180^{th}$ of the temperature difference between the freezing point of water (32°F) and the boiling point of water (212°F) on the Fahrenheit temperature scale.

$$T(°F) = T(°R) - 459.67 = \frac{9}{5} T(°C) + 32$$

Where °R and °C are symbols for degrees Rankin and Celsius, respectively.

Farad (F)

The unit of capacitance in the MKSA system of units equal to the capacitance of a capacitor that has a charge of one Coulomb when a potential difference of one volt is applied.

Faraday's Law (also called the law of electromagnetic induction)

When a magnetic field cuts a conductor, or when a conductor cuts a magnetic field, an electrical current will flow through the conductor if a closed path is provided over which the current can circulate; i.e.,

Differential form: $\text{Curl}\overline{E} = \dfrac{-\partial \overline{B}}{\partial t}$,

Integral form: $\displaystyle\oint_C \overline{E} \cdot \overline{d\ell} = -\int_S \frac{\partial \overline{B}}{\partial t} \cdot \overline{dS} = -\frac{d\Phi}{dt}$

Ferrimagnetics

Ferrimagnetics are materials with the relative permeability (μ_r) much greater than that of vacuum having $\mu_r=1$. Ferrimagnetic materials are materials made of iron oxides (chemical formula: XFe_2O_3, where X is a metal ion), where their internal magnetic moments are not all aligned in one direction, that is to say some are aligned antiparallel, but with smaller magnitudes, so that the net magnetic field output is still much higher than a paramagnetic material. Examples of ferrimagnetics include materials such as manganese-zinc ferrite, barium ferrite, and a whole class of materials, having a high electrical resistance, called ferrites. Ferrimagnetic materials exhibit hysteresis, which is a type of material behavior characterized by an inability to retrace exactly the input-output curve when the magnetizing force is reversed. This nonlinear behavior is caused by the fact that the material will retain some of the magnetic effects internally (called the remnant magnetism) even when the external magnetizing force is completely removed.

Ferromagnetics
ferromagnetics are materials with the relative permeability (μ_r) much greater than that of vacuum ($\mu_r=1$), the amount depending on the magnetizing force. Ferromagnetic materials are a group of materials whose internal magnetic moments align in a common direction such as iron, nickel, cobalt, and their alloys. Ferromagnetic materials exhibit hysteresis, which is a type of material behavior characterized by an inability to retrace exactly the input-output curve when the magnetizing force is reversed. This nonlinear behavior is caused by the fact that the material will retain some of the magnetic effects internally (called the remnant magnetism) even when the external magnetizing force is completely removed.

Field
An entity that acts as an intermediary agent in interactions between particles, is distributed over a region of space, and whose properties are a function of space and time, in general.

Field Theory
The concept that, within a space in the vicinity of a particle, there exists a field containing energy and momentum, and that this field interacts with neighboring particles and their fields.

Flow
The passage of particles (e.g., electrons, etc.) between two points. Example: electrons moving from one terminal of a battery to the other terminal through a conductor. The direction of flows are from higher to lower potential energy levels.

Force
That form of energy that puts an unmoving object into motion, or alters the motion of a moving object (i.e., its speed, direction or both). Furthermore, it is the agency that accomplishes work.

Frequency

The number of complete cycles in one second of a repeating quantity, such as an alternating current, voltage, electromagnetic waves, etc.

G

Gain

The ratio that identifies the increase in signal or amplification that occurs when the signal passes through a circuit.

Ground

(a) A metallic connection with the earth to establish zero potential (used for protection against short circuit); (b) The voltage reference point in a circuit. There may or may not be an actual connection to earth but it is understood that a point in the circuit said to be at ground potential could be connected to earth without disturbing the operation of the circuit in any way.

Gauss

The unit of magnetic induction (also called *magnetic flux density*) in the cgs system of units equal to one line per square centimeter, which is the magnetic flux density of one Maxwell per square centimeter, or 10^{-4} Tesla.

Gauss's Law (electric)

The summation of the normal component of the electrical displacement vector over any closed surface is equal to the electric charges within the surface, which means that the source of the electric flux lines is the electric charge; i.e.,

Differential form: $\mathrm{Div}\overline{D} = \rho$

Integral form: $\oint_S \overline{D} \cdot \overline{dS} = \int_V \rho dv = Q$

Gauss's Law (magnetic)

The summation of the normal component of the magnetic flux density vector over any closed surface is equal to zero, which in essence means that the magnetic flux lines have no source or magnetic charge; i.e.,

Differential form: $\mathrm{Div}\overline{B} = 0$,

Integral form: $\oint_{S} \overline{B} \cdot \overline{dS} = 0$

Generalized Application Mass (G.A.M.)

In general, is any created space, which contains created energies and created matter of any form, shape or size existing as a function of time. In simple terms, generalized application mass is any matter and energy, condensed and packaged into an object form, which exists in a time-stream (from its inception to now). The generalized concept of application mass includes the entire mechanical space containing all energies and matter such as electrons, atoms, molecules and all the existing gigantic masses of planets, stars, galaxies, which are not the direct byproduct of Man's sciences.

Generalized Ohm's Law

When dealing with linear circuits under the influence of time harmonic signals, Ohm's law can be restated under the steady-state condition in the phasor domain as V=ZI, where Z is a complex number called impedance and V and I are voltage and current phasors, respectively.

Gilbert (Gi)

The unit of magnetomotive force in the cgs system of units, equal to the magnetomotive force of a closed loop of one turn in which there is a current of $10/4\pi$ amperes. One Gilbert equals $10/4\pi$ Ampere-turn.

Gradient (of a scalar function)

Gradient (of a scalar function) is a vector, which lies in the direction of maximum rate of increase of the function at any given point and therefore is normal to the constant-value surfaces. Mathematically, it is a vector obtained from a real function f(x,y,z), whose components are the partial derivatives of f(x,y,z), e.g., in Cartesian coordinate system we can write: gradf=($\partial f/\partial x$, $\partial f/\partial y$, $\partial f/\partial z$).

H

Henry (H)
The unit of self and mutual inductance in the MKSA system of units equal to the inductance of a closed loop that gives rise to a magnetic flux of one Weber for each ampere of current that flows through.

Hertz (Hz)
The unit of frequency equal to the number of cycles of a periodic function that occur in one second.

Hole
A vacant electron energy state near the top of the valence band in a semiconductor material. It behaves as a positively charged particle having a certain mass and mobility. It is the dual of electron, unlike a proton which is the dichotomy of an electron.

Hypothesis
An unproven theory or proposition tentatively accepted to explain certain facts or to provide a basis for further investigation.

I

Impedance
The total opposition that a circuit presents to an AC signal, and is a complex number equal to the ratio of the voltage phasor (V) to the current phasor (I).

Incident Wave
A wave that encounters a discontinuity in a medium, or encounters a medium having a different propagation characteristics.

Inductance (L)
The inertial property of an element (caused by an induced reverse voltage), which opposes the flow of current when a voltage is applied; it opposes a change in current that has been established.

Inductor
A conductor used to introduce inductance into an electric circuit, normally configured as a coil to maximize the inductance value.

Input
The current, voltage, power, or other driving force applied to a circuit or device.

Insulator
A material in which the outer electrons are tightly bound to the atom and are not free to move. Thus, there is negligible current through the material when a voltage is applied.

Integrated Circuit (IC)
An electrical network composed of two or more circuit elements on a single semiconductor substrate.

Isolation
Electrical separation between two points.

J

Joule (J)
The unit of energy or work in the MKSA system of units, which is equal to the work performed as the point of application of a force of one Newton moves the object through a distance of one meter in the direction of the force.

Junction
A joining of two different semiconductors or of semiconductor and metal.

Junction Capacitance
The capacitance associated with a junction such as the capacitance of a region of transition between p- and n-type semiconductor materials.

K

Kelvin (K)

The unit of measurement of temperature in the absolute scale (based on Celsius temperature scale), in which the absolute zero is at − 273.15 °C. It is precisely equal to a value of 1/273.15 of the absolute temperature of the triple point of water, being a particular pressure and temperature point, 273.15 K, at which three different phases of water (i.e., vapor, liquid, and ice) can coexist at equilibrium. *See also* **temperature**.

Kinetic

(*Adjective*) Pertaining to motion or change. (*Noun*) Something which is moving or changing constantly such as a piece of matter.

Kinetic Energy (K.E.)

The energy of a particle in motion. The motion of the particle is caused by a force on the particle.

Kirchhoff's Current Law (KCL)

The law of conservation of charge that states that the total current flowing to a given point in a circuit is equal to the total current flowing away from that point.

Kirchhoff's Voltage Law (KVL)

An electrical version of the law of conservation of energy that states that the algebraic sum of the voltage drops in any closed path in a circuit is equal to the algebraic sum of the electromotive forces in that path.

Knowledge

Is a body of facts, principles, data, and conclusions (aligned or unaligned) on a subject, accumulated through years of research and investigation, that provides answers and solutions in that subject.

L

Law
An exact formulation of the operating principle in nature observed to occur with unvarying uniformity under the same conditions.

Law of Conservation of Energy (Excluding All Metaphysical Sources of Energy)
This fundamental law simply states that any form of energy in the physical universe can neither be created nor destroyed, but only converted into another form of energy (also known as the principle of conservation of energy).

Leyden Jar
The first electric capacitor (or condenser) capable of storing charge; it consists of a glass jar with a coat of tin foil outside and inside and a metallic rod passing through the lid and connecting with the inner tin lining. It is named after the city of Leyden (also written as Leiden) in Holland, where it was invented.

Light Waves
Electromagnetic waves in the visible frequency range, which ranges from 400 nm to 770 nm in wavelength.

Linear Network
A network in which the parameters of resistance, inductance, and capacitance of the lumped elements are constant with respect to current or voltage, and in which the voltage or current sources are independent of or directly proportional to other voltages and currents or their derivatives, in the network.

Load
The impedance to which energy is being supplied.

Lossless
A theoretically perfect component that has no loss and hence, transmits all of the energy fed to it.

Lumped Element
A self-contained and localized element that offers one particular electrical property throughout the frequency range of interest.

M

Magnet
A piece of ferromagnetic or ferromagnetic material whose internal domains are sufficiently aligned so that it produces a considerable net magnetic field outside of itself and can experience a net torque when placed in an external magnetic field.

Magnetic Field
The space surrounding a magnetic pole, a current-carrying conductor, or a magnetized body that is permeated by magnetic energy and is capable of exerting a magnetic force. This space can be characterized by magnetic lines of force.

Magnetic Field Intensity (H)
The force that a magnetic field would exert on a unit magnetic pole placed at a point of interest, which expresses the free space strength of the magnetic field at that point (also called *magnetic field strength*, *magnetic intensity*, *magnetic field*, *magnetic force*, and *magnetizing force*).

Magnetostatics
The study of magnetic fields that are neither moving nor changing direction.

Man
Homo sapiens (literally, the knowing or intelligent man); mankind.

Mathematics
Mathematics are short-hand methods of stating, analyzing, or resolving real or abstract problems and expressing their solutions by symbolizing data, decisions, conclusions, and assumptions.

Matter
Matter particles are a condensation of energy particles into a very small volume.

Maxwell (Mx)
The unit for magnetic flux in the cgs system of units, equal to 10^{-8} Weber.

Maxwell's Equations

A series of four advanced classical equations developed by James Clerk Maxwell between 1864 and 1873, which describe the behavior of electromagnetic fields and waves in all practical situations. They relate the vector quantities for electric and magnetic fields as well as electric charges existing (at any point or in a volume), and set forth stringent requirements that the fields must satisfy. These celebrated equations are given as follows:

Differential form   Integral form

1) Ampere's Law: $\mathrm{Curl}\overline{H} = \overline{J} + \dfrac{\partial \overline{D}}{\partial t}$, $\oint_C \overline{H} \cdot \overline{d\ell} = I + \int_S \dfrac{\partial \overline{D}}{\partial t} \cdot \overline{dS}$

2) Faraday's Law: $\mathrm{Curl}\overline{E} = \dfrac{-\partial \overline{B}}{\partial t}$, $\oint_C \overline{E} \cdot \overline{d\ell} = -\int_S \dfrac{\partial \overline{B}}{\partial t} \cdot \overline{dS} = -\dfrac{d\Phi}{dt}$

3) Gauss's Law (electric): $\mathrm{Div}\overline{D} = \rho$, $\oint_S \overline{D} \cdot \overline{dS} = \int_V \rho dv = Q$

4) Gauss's Law (magnetic): $\mathrm{Div}\overline{B} = 0$, $\oint_S \overline{B} \cdot \overline{dS} = 0$

From these equations, Maxwell predicted the existence of electromagnetic waves whose later discovery made radio possible. He showed that where a varying electric field exists, it is accompanied by a varying magnetic field induced at right angles, and vice versa, and the two form an electromagnetic field pair that could propagate as a transverse wave. He calculated that in a vacuum, the speed of the wave was given by $1/\sqrt{(\varepsilon_0\mu_0)}$, where ε_0 and μ_0 are the permittivity and permeability of vacuum. The calculated value for this speed was in remarkable agreement with the measured speed of light, and Maxwell concluded that light is propagated as electromagnetic waves.

Mechanics

The totality of the three categories of application mass: a) Generalized application Mass; b) Technical application mass, and c) Personalized application mass. See also classical mechanics and quantum mechanics.

Microelectronics

The body of electronics that is associated with or applied to the realization of electronic systems from extremely small electronic parts.

Microstrip Line

A microwave transmission line that is composed of a single conductor supported above a ground plane by a dielectric.

Microwave Integrated Circuit (MIC)

A circuit that consists of an assembly of different circuit functions that are connected by Microstrip transmission lines. These different circuits all incorporate planar semiconductor devices, passive lumped elements, and distributed elements.

Microwaves

Waves in the frequency range of 1 GHz to 300 GHz.

Millimeter Wave

Electromagnetic radiation in the frequency range of 30 to 300 GHz, corresponding to wavelength ranging from 10 mm to 1 mm.

Model

A physical (e.g., a small working replica), abstract (e.g., a procedure) or a mathematical representation (e.g., a formula) of a process, a device, a circuit, or a system and is employed to facilitate their analysis.

Monolithic Integrated Circuit

An integrated circuit that is formed in a single block or wafer of semiconductor materials. The term is derived from Greek, "monolithos", which means "made of one stone."

Monolithic Circuits

Are integrated circuits entirely on a single chip of semiconductor.

Monolithic Microwave Integrated Circuit (MMIC)

A microwave circuit obtained through a multilevel process approach comprising of all active and passive circuit elements as well as interconnecting transmission lines, which are formed into the bulk or onto the surface of a semi-insulating semiconductor substrate by some deposition scheme such as epitaxy, ion implantation, sputtering, evaporation, diffusion, etc.

Monumental Discovery
Any of the six major un-ravelings or breakthroughs of knowledge about a significant phenomenon in the field of electricity, which shifted the subject in a substantial way and expanded all of the hitherto knowledge amply.

N

Natural Laws
A body of workable principles considered as derived solely from reason and study of nature.

Neper (Np)
A unit of attenuation used for expressing the ratio of two currents , voltages, or fields by taking the natural logarithm (logarithm to base e) of this ratio. If voltage V_1 is attenuated to V_2 so that $V_2/V_1 = e^{-N}$, then N is attenuation in Nepers (always a positive number) and is defined by: $N \, (Np) = \log_e(V_1/V_2) = \ln(V_1/V_2)$, where $V_1 > V_2$.

Neutron
One of uncharged stable elementary particles of an atom having the same mass as a proton. A free neutron decomposes into a proton, an electron, and a neutrino. A neutrino is a neutral uncharged particle but is an unstable particle since it has a mass that approaches zero very rapidly (a half-life of about 13 minutes).

Network
A collection of electric devices and elements (such as resistors, capacitors, etc.) connected together to form several interrelated circuits.

Newton (N)
The unit of force in MKSA system of units equal to the force that imparts an acceleration of one m/s^2 to a mass of one kilogram.

Noise
Random unwanted electrical signals that cause unwanted and false output signals in a circuit.

Nomenclature
The set of names used in a specific activity or branch of learning; terminology.

Nonlinear
Having an output that does not rise and fall in direct proportion to the input.

Nucleus
The core of an atom composed of protons and neutrons, having a positive charge equal to the charge of the number of protons that it contains. The nucleus contains most of the mass of the atom, pretty much like the sun containing most of the mass of the solar system.

O

Occam's (or Ockham's) Razor Doctrine
A principle that assumptions introduced to explain a thing must not be multiplied beyond necessity. In simple terms, it is a principle stating that the simplest explanation of a phenomenon, which relates all of the facts, is the most valid one. Thus by using the Occam's razor doctrine a complicated problem can be solved through the use of simple explanations, much like a razor cutting away all undue complexities (after William of Occam, an English philosopher, 1300-1349, who made a great effort to simplify scholasticism).

Oersted (Oe)
The unit of magnetic field in the cgs system of units equal to the field strength at the center of a plane circular coil of one turn and 1-cm radius when there is a current of $10/2\pi$ ampere in the coil.

Ohm (Ω)
The unit of resistance in the MKSA system of units equal to the resistance between two points on a conductor through which a current of one ampere flows as a result of a potential difference of one volt applied between the two points.

Ohm's Law
The potential difference V across the resistor terminals is directly proportional to the electrical current flowing through the resistor. The proportionality constant is called resistance (R); i.e., V=RI. Ohm's Law can also be interpreted as the conversion of potential energy (V) into kinetic energy (I), which is a simple statement expressing the principle of conservation of energy.

Original Postulates
A series of exact postulate (space, energy, change) that have gone into the construction of the physical universe. See primary postulates.

Oscillator
An electronic device that generates alternating-current power at a frequency determined by constants in its circuits.

Output
The current, voltage, power, or driving force delivered by a circuit or device.

P

Paramagnetics
are materials (such as aluminum, beryllium, etc.), which accept magnetism.

Particle
Any tiny piece of matter, so small as to be considered theoretically without magnitude (i.e., zero size), though having mass, inertia and the force of attraction. Knowing zero size is an absolute and thus impossible in the physical universe, practical particles range in diameter from a fraction of angstrom (as with electrons, atoms and molecules) to a few millimeters (as with large rain drops).

Passive
A component that may control but does not create or amplify electrical energy.

Perfect Conductor
Is a conductor having infinite conductivity or zero resistivity.

Personalized Application Mass (P.A.M.)
Is the category of application mass, which has been created and is based solely upon the viewpoint's own postulates and considerations. Examples of this category include such things as one's own customized possessions, any piece of artwork or music, one's own body characteristics (such as hairdo, clothing, shape, etc.), a book's layout or cover design, so on and so forth. see also application mass, Technical application mass, and Generalized application mass.

Phase
The angular relationship of a wave to some time reference or other wave.

Phase Constant
The imaginary component of the propagation constant for a traveling wave at a given frequency.

Phasor
A result of a mathematical transformation of a sinusoidal waveform (voltage, current, or EM wave) from the time domain into the complex number domain (or frequency domain) whereby the magnitude and phase angle information of the sinusoid is retained.

Physical Universe (Also Called Material Universe; The Universe)
Is a universe based upon three postulates, called original postulates (space, energy and change) and has four main components (matter, energy, space and time).

Plane Wave
A wave whose wave fronts are plane surfaces and normal to the direction of propagation.

Plating
See electroplating.

PN Junction
An abrupt transition between p-type and n-type semiconductor regions within a crystal lattice. Such a junction possesses specific electrical properties such as the ability to conduct in only one direction, and is used as the basis for semiconductor devices, such as diodes, transistors, etc.

Port
Access point to a system or circuit.

Postulate
a) (NOUN) is an assumption or assertion set forth and assumed to be true unconditionally and for all times without requiring proof; especially as a basis for reasoning or future scientific development; b) (VERB) To put forth or assume a datum as true or exist without proof.

Potential Difference (or Voltage)
The electrical pressure or force between any two points caused by accumulation of charges at one point relative to another, which has the capability of creating a current between the two points.

Potential Energy (P.E.)
Any form of stored energy that has the capability of performing work when released. This energy is due to the position of particles relative to each other.

Power
The rate at which work is performed; i.e., the rate at which energy is being either generated or absorbed.

Primary Postulates
A series of four postulates derived from original postulates. These postulates are responsible for the four basic components of the physical universe: matter, energy, created space, and mechanical time. See original postulates.

Principle
A rule or law illustrating a natural phenomenon, operation of a machine, the working of a system, etc.

Processing
The act of converting material from one form into another more desired form, such as in integrated circuit fabrication where one starts with a wafer and through many steps ends up with a functional circuit on a chip.

Propagation
The travel of electromagnetic waves through a medium.

Propagation Constant
A number showing the effect (such as losses, wave velocity, etc.) a transmission line has on a wave as it propagates along the line. It is a complex term having a real term called the *attenuation factor* and an imaginary term called the *phase constant*.

Proton
An elementary particle, which is one of the three basic subatomic particles, with a positive charge equivalent to the charge of an electron (q= +1.602x10^{-19} C) and has a mass of about 1.67x10^{-27} kg with a spin of ½. Proton together with neutron is the building block of all atomic nuclei.

Pulse
A variation of a quantity, which is characterized by a rise to a certain level (amplitude), a finite duration, and a decay back to the normal level.

Pyramid of Knowledge
Workable knowledge forms a pyramid, where from a handful of common denominators efficiently expressed by a series of basic postulates, axioms and natural laws, which form the foundation of a science, an almost innumerable number of devices, circuits and systems can be thought up and developed. The plethora of the mass of devices, circuits and systems generated is known as the "application mass", which practically approaches infinity in sheer number.

Q

Quantum Mechanics (Also Called Quantum Physics or Quantum Theory)
Is the study of atomic structure which states that an atom or molecule does not radiate or absorb energy continuously. Rather, it does so in a *series of steps, each step being the emission or absorption of an amount of energy packet (E) called a quantum.* Quantum physics is the modern theory of matter, electromagnetic radiation and their interaction with each other. It differs from

classical physics in that it generalizes and supersedes it, mainly in the realm of atomic and subatomic phenomena.

Quark
A hypothetical basic particle having a fraction of charge of an electron (such as 1/3 or 2/3) from which many of the elementary particles (such as electrons, protons, neutrons, mesons, etc.) may be built up theoretically. No experimental evidence for the actual existence of free quarks has been found.

R

Radio Frequency (RF)
Any wave in the frequency range of a few kHz to 300 MHz, at which coherent electromagnetic radiation of energy is possible.

Rankine (°R)
The unit of measurement of temperature in the absolute scale (based on Fahrenheit temperature scale), in which the absolute zero is at -459.67 °F. *See also* **temperature**.

Reactance
Is a parameter that is the measure of the opposition to the flow of alternating current (Symbolized by X).

Reactive Element
Is an element, which impedes the flow of current in a wire. An inductor or a capacitor are reactive elements. A purely reactive element does not dissipate energy as does a resistor, but stores it in the associated electric and/or magnetic fields.

Rectifier
Is a device having an asymmetrical conduction characteristic such that current can flow in only one direction through the device.

Reflected Waves
The waves reflected from a discontinuity back into the original medium, in which they are traveling.

Reflection Coefficient
The ratio of the reflected wave phasor to the incident wave phasor.

Resistance
A property of a resistive material that determines the amount of current flow when a voltage is applied across it. The resistor value is dependent upon geometrical dimensions, material, and temperature.

Resistor
A lumped bilateral and linear element that impedes the flow of current, i(t), through it when a potential difference, V (t), is imposed between its two terminals. The resistor's value is found by: $R=V(t)/i(t)$.

Resonant Frequency
The frequency at which a given system or circuit will respond with maximum amplitude when driven by an external sinusoidal force.

Right-Hand Rule
For a current-carrying wire, the rule that if the fingers of the right hand are placed around the wire so that the thumb points in the direction of the current flow, the finger curling around the wire will be pointing in the direction of the magnetic field produced by the wire.

S

Science
A branch of study concerned with establishing, systematizing, and aligning laws, facts, principles, and methods that are derived from hypothesis, observation, study and experiments.

Semiconductor
A material having a resistance between that of conductors and insulators, and usually having a negative temperature coefficient of resistance.

Signal
An electrical quantity (such as a current or voltage) that can be used to convey information for communication, control, etc.

Silicon (Si)
A semiconductor material element in column IV of the periodic table used as in device fabrication.

Sinusoidal
Varying in proportion to the sine or cosine of an angle or time function. For example, the ordinary AC signal is a sinusoidal.

Small Signal
A low-amplitude signal that covers such a small part of the operating characteristic curve of a device that operation is nearly always linear.

Solid-State Device
Any element that can control current without moving parts, heated filaments, or vacuum gaps. All semiconductors are solid-state devices, although not all solid-state devices (such as transformers, ferrite circulators, etc.) are semiconductors.

Space (Also Called Created Space)
The continuous three-dimensional expanse extending in all directions, within which all things under consideration exist.

Standing Wave
A standing, apparent motionless-ness, of particles causing an apparent no out-flow, no in-flow. A standing wave is caused by two energy flows, impinging against one another, with comparable magnitudes to cause a suspension of energy particles in space, enduring with a duration longer than the duration of the flows themselves.

Standing Wave Ratio (SWR)
The ratio of current or voltage on a transmission line that results from two waves having the same frequency and traveling in opposite directions meeting and creating a standing wave.

Static
(Adjective) Pertaining to no-motion or no-change. (Noun) Something which is without motion or change such as truth (an abstract concept). In physics, one may consider a very distant star (a physical universe object) a static on a short term basis, but it is not totally correct because the distant star is moving over a long period of time, thus is not truly a static but only an approximation, or a physical analogue of a true static.

Subjective Time
Is the consideration of time in one's mind, which can be a nonlinear or linear quantity depending on one's viewpoint.

Substrate
A single body of material on or in which one or more electronic circuit elements or integrated circuits are fabricated.

Superposition Theorem
This theorem states that in a linear network, the voltage or current in any element resulting from several sources acting together is the sum of the voltages or currents resulting from each source acting alone, while all other independent sources are set to zero; i.e., $f(v_1+v_2+\ldots\ldots\ldots\ldots+v_n)=f(v_1)+f(v_2)+\ldots\ldots\ldots\ldots+f(v_n)$

Supplemental Discovery
Any of the eight subordinate discoveries (along with their magnetic duals), which fill in the gaps left behind by the six monumental discoveries of electricity.

Switch
A mechanical or electrical device that completes or breaks the path of the current or sends it over a different path.

Switching
Is the making, breaking, or changing of connections in an electronic or electric circuit.

Symbiont
An organism living in a state of association and interdependence with another kind of organism, especially where such association is of mutual advantage, such as a pet. Such a state of mutual interdependence is called "symbiosis."

T

Technical Application Mass (T.A.M.)
Is the category of man-made application mass that is produced directly as a result of application of a science using its scientific postulates, axioms, laws and other technical data. Examples include such things as a television set, a computer, an automobile, a power

generator, a telephone system, a rocket, etc. See also Application mass, Personalized application mass, and Generalized application mass.

Technology
The application of a science for practical ends.

Temperature
The degree of hotness or coldness measured with respect to an arbitrary zero or an absolute zero, and expressed on a degree scale. Examples of arbitrary-zero degree scales are Celsius scale (°C) and Fahrenheit scale (°F); and examples of absolute-zero degree scales are Kelvin degree scale (based on Celsius degree scale) and Rankine degree scale (based on Fahrenheit degree scale).

Tesla (T)
The unit of magnetic field in the MKSA system of units equal to one Weber per square meter.

TEM (Transverse Electro-Magnetic) Wave
Waves having the electric and magnetic fields perpendicular to each other and to the direction of propagation. These waves have no field components in the direction of propagation.

Theorem
A proposition that is not self-evident but can be proven from accepted premises and therefore, is established as a principle.

Theory
An explanation based on observation and reasoning, which explains the operation and mechanics of a certain phenomenon. It is a generalization reached by inference from observed particulars and implies a larger body of tested evidence and thus a greater degree of probability. It uses a hypothesis as a basis or guide for its observation and further development.

Thermal Noise (Johnson Noise or Nyquist Noise)
The most basic type of noise that is caused by thermal vibration of bound charges and thermal agitation of electrons in a conductive material. This is common to all passive or active devices.

Time (Also Called Mechanical Time or Objective Time)

That characteristic of the physical universe at a given location that orders the sequence of events on a microscopic or macroscopic level. It proceeds from the interaction of matter and energy and is merely an "index of change," used to keep track of a particle's location. The fundamental unit of time measurement is supplied by the earth's rotation on its axis while orbiting around the sun. It can also alternately be defined as the co-motion and co-action of moving particles relative to one another in space. See also subjective time.

Torque

A force that tends to produce rotation or twisting.

Transformer

An electrical device that, by electromagnetic induction, transforms electric energy from one (or more) circuit(s) to one (or more) other circuit(s) at the same frequency, but usually at a different voltage and current value.

Transmission Line (T.L.)

Any system of conductors suitable for conducting electric or electromagnetic energy efficiently between two or more terminals.

Transmitted wave

That portion of an incident wave that is not reflected at the interface, but actually travels from one medium to another.

Two-port network

A network that has only two access ports, one for input or excitation, and one for output or response.

U

Unidirectional

Flowing in only one direction (e.g., direct current).

Unilateral

Flowing or acting in one direction only causing a non-reciprocal characteristic.

Universal Communication Principle (Also Called Communication Principle)

A fundamental concept in life and livingness that is intertwined throughout the entire field of sciences that states for communication to take place between two or more entities, three elements must be present: a source point, a receipt point, and an imposed space or distance between the two.

Universe (Derived From Latin Meaning "Turned Into One", "A Whole)

Is the totality or the set of all things that exist in an area under consideration, at any one time. In simple terms, it is an area consisting of things (such as ideas, masses, symbols, etc.) that can be classified under one heading and be regarded as one whole thing.

Viewpoint

Is a point on a mental plane from which one creates (called postulating viewpoint) or observes (called observing viewpoint) an idea, an intended subject or a physical object.

Volt (V)

The unit of potential difference (or electromotive force) in the MKSA system of units equal to the potential difference between two points for which one Coulomb of charge will do one joule of work in going from one point to the other.

Voltage

Voltage or potential difference between two points is defined to be the amount of work done against an electric field in order to move a unit charge from one point to the other.

Voltage Source

The device or generator connected to the input of a network or circuit.

Voltage Standing Wave Ratio (VSWR)

The ratio of maximum voltage to the minimum voltage on a transmission. The standing wave on a line results from two voltage (or current) waves having the same frequency, and traveling in opposite directions.

W

Wafer
A thin semiconductor slice of silicon or germanium on which matrices of microcircuits or individual semiconductors can be formed using manufacturing processes. After processing, the wafer is separated into chips (or *die*) containing individual circuits.

Watt (W)
The unit of power in MKSA system of units defined as the work of one joule done in one second.

Wave
A disturbance that propagates from one point in a medium to other points without giving the medium as a whole any permanent displacement.

Wave Propagation
The travel of waves (e.g., electromagnetic waves) through a medium.

Waveguide
A transmission line comprised of a hollow conducting tube within which electromagnetic waves are propagated.

Wavelength
The physical distance between two points having the same phase in two consecutive cycles of a periodic wave along a line in the direction of propagation.

Weber (Wb)
The unit of magnetic flux in the MKSA system of units equal to the magnetic flux, which linking a circuit of one turn, produces an electromotive force of one volt when the flux is reduced to zero at a uniform rate in one second.

Work
The advancement of the point of application of a force on a particle.

Technical References

1. Boas, M. *The Scientific Renaissance,* New York: Harper, 1965.
2. Bordeau, S. B. *Volts to Hertz ... The Rise of Electricity, Minneapolis*: Burgess Publishing Co., 1982.
3. Bowick, C. *RF Circuit Design*, Upper Saddle River: SAMS-Prentice Hall, 1982.
4. Buckley, H. *A Short History of Physics*, London: Methuen, 1927.
5. Burns, S.B. and P. R. Bond, *Principles of Electronic Circuits*, St. Paul: West, 1987.
6. Cajori, F., *A History of Physics*, New York: Dover, 1962.
7. Carr, J. J., *Secrets of RF Circuit Design*, New York: McGraw-Hill, 1991.
8. Chaffin, R. J., *Microwave semiconductor Devices*, New York: Wiley & Sons, 1973.
9. Cheng, D. K., *Fundamentals of Engineering Electromagnetics*, Reading: Addison Wesley, 1993.
10. Chang, K. *Microwave Solid-State Circuits and Applications*, New York: John Wiley & Sons, 1994.
11. Chang K. Microwave and Optical Component, Vols. I and II, New York: Wiley, 1989.
12. Cheung, W. S. and F. H., Levien. *Microwave made simple.* Dedham: Artech House, 1985.
13. Collin, R. E. *Foundation for Microwave Engineering*, 2nd Ed., New York: McGraw-Hill, 1992.
14. Dampier, W. C. *A History of Science,* Cambridge: Cambridge University Press, 1966.
15. Desor, C. A. and E. S., Kuh *Basic Circuit Theory*, Tokyo: McGraw-Hill, 1969.

16. Dorf, R. C. and J. A. Savoboda, *Introduction to Electric Circuits*, New York: Wiley & Sons, 1996.

17. Dorf, R. C. *Electrical Engineering Handbook*, Boca Raton: CRC press. FL, 1993

18. Edwards, T. C. *Foundations for Microstrip Circuit Design*, New York: Wiley & Sons, 1981.

19. Flanagan, W. N. *Handbook of Transformer Design and Application*, Blue Ridge Summit: McGraw-Hill, 1993.

20. Fortino, A. G. *Fundamentals of Integrated circuit technology*, Reston Publishing Company (Prentice-Hall), 1984.

21. Gardiol, F. E. *Introduction to Microwaves*, Dedham: Artech House, 1984.

22. Gardiol, F. *Microstrip Circuits*, New York: Wiley, 1994.

23. Gibilisco, S. *The Illustrated Dictionary of Electronics*, Blue Ridge Summit: Tab Books, 1994.

24. Gillmore, C. S. *Coulomb and the Evolution of Physics and Engineering in the 18th Century,* Princeton, NJ: Princeton University Press, 1971.

25. Gonzalez, G., *Microwave Transistor Amplifiers, Analysis and Design*, 2nd ed. Upper Saddle River: Prentice Hall, 1997.

26. Gottlieb, I. W., *Practical RF Power Design Techniques*, New York: McGraw-Hill, 1993.

27. Gray, P. R., and C. L. Searle, *Electronic Principles*, 3rd edition, New York: Wiley 1993.

28. Ha, T. T., *Solid State Microwave Amplifier Design*. New York: Wiley & Sons, 1987.

29. Hambley, A. R., *Electronics*, 2nd Edition, Prentice Hall, 2000.

30. Hamilton, D. J., W. G. Howard. *Basic Integrated Engineering*, New York: McGraw-Hill, 1975.

31. Haus, H. A., and R. B. Adler. *Circuit Theory of Linear Noisy Networks*, Cambridge, MA, MIT Press, 1959 and New York: Wiley & Sons, 1959.

32. Harsang, S. C. *Principles of Microwave Technology*, Upper Saddle River: Prentice Hall, 1997.

33. Hart, I. *Makers of Science*, New York: Oxford University Press, 1923.

34. Hawking, S. W. *A Brief History of Time*, New York: Bantam Books, 1988.

35. Hodges, D.A., and H.G. Jackson. *Analysis and Design of Digital Integrated Circuits*, 2nd Edition, New York: McGraw-Hill, 1988.

36. Horowitz, P., and W. Hill. *The Art of Electronics*, Cambridge University Press, 1983.

37a. Hubbard, L. Ron. *Understanding the E-meter*, Bridge Publications, 1982.

37b. Hubbard, L. Ron. *To the Stars*, Galaxy Press, 2004.

38. Ishii, T. K. *Microwave Engineering*, 2nd ed., Orlando: Harcourt Brace Jovanovich, publishers, 1989.

39. Johns, A., and S. E. Butner. *Analog Integrated Circuit Design*, New York: Wiley, 1997.

40. Krauss, H. L., C. W. Bostian, and F. H. Raab. *Solid State Radio Engineering,* New York: Wiley & Sons, 1980.

41. Kraus, J. D. *Electromagnetics*, 3rd Ed., New York: McGraw-Hill, 1984.

42. Lackey, J. E., J. L. Massey and M. D. Hehn. *Solid State Electronics*, Holt, Rinehart and Winston, 1986

43. Lapedes, D. N., McGraw-Hill *Dictionary of Physics and Mathematics*, New York: McGraw-Hill, 1978.

44. Laverghetta, T., *Practical Microwaves*, Indianapolis: Howard Sams, 1984.

45. Lenk, J. D. *Lenk's RF Handbook*, New York: McGraw-Hill, 1992.

46. Liao, S. Y. *Microwave Circuit Analysis and Amplifier Design*, Upper Saddle River: Prentice Hall, 1987.

47. Lincoln, E. S. *A Chronological History of Electrical Developments*, National Electrical Manufacturer's Association, New York, 1964.

48. Lipson, H. *Great Experiments in Physics*, Edinburgh: Oliver & Boyd, 1965.

49. Maas, S. A. *Nonlinear Microwave Circuits*, Norwood, MA: Artech House, 1988.

50. Mackenzie, A. E. E. *The Major Advancements in Science,* Cambridge: Cambridge University Press, 1960.

51. Marion, J. V. *A Universe of Physics*, New York: Wiley, 1970.

52. Matthaei, G., L., Young, and E. M. Jones. *Microwave Filters, Impedance-Matching Networks, and Coupling Structures*, Dedham: Artech House, 1980.

53. Meyer, H. W. *A. History of Electricity and Magnetism*, Cambridge, MA: MIT Press, 1972.

54. Millman, J. A., and A. Grabel. *Microelectronics*, 2nd ed. New York: McGraw-Hill, 1987.

55. Navon, H. *Semiconductor Microdevices and Materials*, New York: Holt, Rinehart and Winston, 1986

56. Pledge, H. T. *Science Since 1500.* London: Science Museum, 1966.

57. Plonsey, R. and R. E. Collin. *Principles and applications of Electromagnetic Fields*, 2nd Ed., New York: McGraw-Hill, 1982.

58. Pozar, D. M. *Microwave Engineering*, 2nd ed. New York: Wiley & Sons, 1998.

59. Radmanesh, M. M. *Radio Frequency and Microwave electronics Illustrated*, Upper Saddle River: Prentice Hall, 2001.

60. Radmanesh, M. M. *Obstacles to Comprehension of Engineering Sciences*, Annual IEE Engineering Conference Digest, Sacramento, CA, April 1992.

61. Radmanesh, M. M. *Creativity in Engineering Education for Higher Student Retention*, ASEE Pacific Southwest (PSW) Engineering Conference Digest, Flagstaff, AZ, October 1993.

62. Runes, D. D. *A Treasury of World Science*, New York: Philosophical Library, 1961.

63. Saad, T. *Microwave Engineer's Handbook*, Vols. I, II. Dedham: Artech House, 1988.

64. Schwartz, G., and P. W. Bishop. *Moments of Discovery*, New York: Basic Books, 1962.

65. Scott, A. W. *Understanding Microwaves*, New York: Wiley & Sons, 1993.

66. Sedra, A. S., and K. C. Smith. *Microelectronic Circuits*, New York: Oxford University Press, 1998.

67. Shen, L. C., and J. A. Kong. *Applied Electromagnetism*, 2nd ed., Boston: PWS Engineering, 1987.

68. Silsbee, F. B. *Systems of Electrical Units*, U.S. Government printing office: National Bureau of Standards, Monograph 56, 1962.

69. Soclof, S. *Analog Integrated Circuits*, Englewood cliffs, NJ: Prentice-Hall 1985.

70. Streetman, B.J. *Solid State Electronic Devices*, Third Edition, Englewood Cliffs, NJ: Prentice-Hall, 2000.
71. Taton, R. *History of Science*, Vols. I-IV. New York: Basic Books, 1964.
72. Tricker, R. A. R. *The Contributions of Faraday and Maxwell to Electrical Science*, Pergamon Press, New York, 1966.
73. Vendelin, G. D., *Design of Amplifiers and Oscillator by the S-Parameter Method,* New York: Wiley, 1981.
74. Vendelin, G. D., A. M. Pavio, and Ulrich L. Rhode. *Microwave Circuit Design,* New York: Wiley & Sons, 1990.
75. Vizmuller,P. *RF Design Guide*, Norwood: Artech House, 1995.
76. Wadell, B. C. *Transmission Line Design Handbook*, Norwood: Artech House, 1991.
77. Watson, H. A. *Microwave Semiconductor Devices and Their Circuit Applicatio*ns, New York: McGraw-Hill, 1969.
78. White, J. F. *Microwave Semiconductor Engineering*, Van Norstrand, Princeton, N. J., 1990.
79. Wightman, W.P.D. *The Growth of Scientific Ideas*, Yale University press, New Haven, Conn., 1969.
80. Wolff, A. *A history of Science, Technology and Philosophy in the 18th Century*, London: Ruskin House, 1952.
81. Zahn, M. *Electromagnetic Field Theory*, New York: John Wiley & Sons, 1979.

About the Author

Matthew M. Radmanesh received his BSEE degree from Pahlavi University in electrical engineering in 1978, his MSEE and Ph.D. degrees from the University of Michigan, Ann Arbor, in Microwave Electronics and Electro-Optics in 1980 and 1984, respectively.

He has worked in academia for Kettering University (formerly GMI Engineering & Management) and in industry for Hughes Aircraft Co., Maury Microwave Corp. and Boeing Aircraft Co. He is currently a faculty member in the electrical and computer engineering department at California State University, Northridge, CA.

Dr. Radmanesh is a senior member of IEEE, Eta Kappa Nu Honor society and a past president (three years) of the SFV chapter of the IEEE Microwave Theory and Technique (MTT) society. His many years of experience in both microwave industry and academia have led to over 40 technical papers in national and international journals and several design handbooks in RF/Microwave engineering and in solid state devices and integrated circuit engineering.

His current research interests include design of RF and Microwave devices and circuits, millimeter-wave circuit applications, photonic engineering as well as engineering education. He received the distinguished lecturer award at the 1994 IEEE international Microwave Symposium and was awarded twice by IEEE LA council for his contributions to the MTT society (1994, 1995). He also received two awards for commitment and dedication to education from IEEE, in 2002 and 2003.

Dr. Radmanesh won the MPD divisional award while at Hughes Aircraft Co. for his pioneering work in the development and design of solid state millimeter wave noise sources in Ka- and V-band, and a similar award for his outstanding contributions to the HERF project from Boeing Aircraft Co.

Dr. Radmanesh authored a popular textbook entitled "Radio Frequency and Microwave Electronics Illustrated," published in U.S. and internationally by Prentice Hall in 2001. The Chinese translation of this work was published in the People's Republic of China in 2002.

He holds two patents for his pioneering work and novel designs of two millimeter-wave noise sources. His hobbies include chess, philosophy, soccer, and tennis.

Printed in the United States
66314LVS00003B/125

9 781420 839999